Circuit Analysis

FOR

DUMMIES®

A Wiley Brand

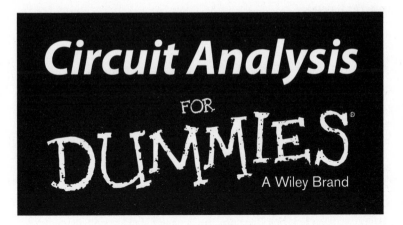

by John M. Santiago, Jr., PhD

Professor of Electrical and Systems
Engineering, Colonel (Ret) USAF

Circuit Analysis For Dummies®

Published by
John Wiley & Sons, Inc.
111 River St.
Hoboken, NJ 07030-5774
www.wiley.com

Copyright © 2013 by John Wiley & Sons, Inc., Hoboken, New Jersey

Published by John Wiley & Sons, Inc., Hoboken, New Jersey

Published simultaneously in Canada

No part of this publication may be reproduced, stored in a retrieval system or transmitted in any form or by any means, electronic, mechanical, photocopying, recording, scanning or otherwise, except as permitted under Sections 107 or 108 of the 1976 United States Copyright Act, without the prior written permission of the Publisher. Requests to the Publisher for permission should be addressed to the Permissions Department, John Wiley & Sons, Inc., 111 River Street, Hoboken, NJ 07030, (201) 748-6011, fax (201) 748-6008, or online at http://www.wiley.com/go/permissions.

Trademarks: Wiley, the Wiley logo, For Dummies, the Dummies Man logo, A Reference for the Rest of Us!, The Dummies Way, Dummies Daily, The Fun and Easy Way, Dummies.com, Making Everything Easier, and related trade dress are trademarks or registered trademarks of John Wiley & Sons, Inc., and/or its affiliates in the United States and other countries, and may not be used without written permission. All other trademarks are the property of their respective owners. John Wiley & Sons, Inc., is not associated with any product or vendor mentioned in this book.

For general information on our other products and services, please contact our Customer Care Department within the U.S. at 877-762-2974, outside the U.S. at 317-572-3993, or fax 317-572-4002.

For technical support, please visit www.wiley.com/techsupport.

Wiley publishes in a variety of print and electronic formats and by print-on-demand. Some material included with standard print versions of this book may not be included in e-books or in print-on-demand. If this book refers to media such as a CD or DVD that is not included in the version you purchased, you may download this material at http://booksupport.wiley.com. For more information about Wiley products, visit www.wiley.com.

Library of Congress Control Number: 2013932101

ISBN 978-1-118-49312-0 (pbk); ISBN 978-1-118-59050-8 (ebk); ISBN 978-1-118-59052-2 (ebk); ISBN 978-1-118-59056-0 (ebk)

Manufactured in the United States of America

10 9 8 7 6 5 4 3 2 1

About the Author

John Santiago retired from the military in 2003 with 26 years of service in the United States Air Force (USAF). John has served in a variety of leadership positions in technical program management, acquisition development, and operation research support. While assigned in Europe for three years with the USAF, he spearheaded more than 40 international scientific and engineering conferences/workshops as a steering committee member.

John has experience in many engineering disciplines and missions, including control and modeling of large, flexible space structures; communications systems; electro-optics; high-energy lasers; missile seekers/sensors for precision-guided munitions; image processing/recognition; information technologies; space, air, and missile warning; missile defense; and homeland defense.

One of John's favorite assignments was serving as an associate professor at the USAF Academy during his tour from 1984 through 1989. John is currently a professor of Electrical and Systems Engineering at Colorado Technical University, where he has taught 26 different undergraduate and graduate courses in electrical and systems engineering.

Some of his awards include Faculty of the Year at Colorado Technical University in 2008; USAF Academy Outstanding Military Educator in 1989; and USAF Academy Outstanding Electrical Engineering Educator in 1998.

During his USAF career, John received his PhD in Electrical Engineering from the University of New Mexico; his Master of Science in Resource Strategy at the Industrial College of the Armed Forces; his Master of Science in Electrical Engineering from the Air Force Institute of Technology, specializing in electro-optics; and his Bachelor of Science from the University of California, Los Angeles.

On February 14, 1982, John married Emerenciana F. Manaois.

More information about John's background and experience is available at www.FreedomUniversity.TV.

Dedication

To my heavenly Father, thank you for all the many blessings, especially the gift of family and friends.

To my lovely Emily, thank you for your loving and continued support, always and forever.

To my parents, who bravely immigrated here from the Philippines to live in this great nation.

To the Founding Fathers, who were engineers and visionary leaders in creating this great country called the United States. To their creative genius and to all those standing on their shoulders, especially the next generation of engineers.

To all those who wondered if there's anything more to circuit analysis than Ohm's law and Kirchhoff's laws.

Author's Acknowledgments

Many people have been involved in this book, and I thank them all. First, to all the former cadets at the United States Air Force Academy, the students at the University of West Florida, and the students at the Colorado Technical University who endured my class lessons and asked a gazillion questions over the years as I continued to learn to become a better teacher.

I'd like to thank a team of players who've made my writing presentable. First, I'd like to thank Matt Wagner, my agent, who contacted me about writing this book. I'd also especially like to thank the folks at Wiley who made this book possible: Erin Mooney, my acquisitions editor; Jennifer Tebbe, my project editor; and Danielle Voirol, my copy editor.

I'd again like to thank my wife for her encouragement and for keeping me straight about the things going on at home and in the community as I went through this writing adventure.

Publisher's Acknowledgments

We're proud of this book; please send us your comments at http://dummies.custhelp.com. For other comments, please contact our Customer Care Department within the U.S. at 877-762-2974, outside the U.S. at 317-572-3993, or fax 317-572-4002.

Some of the people who helped bring this book to market include the following:

Acquisitions, Editorial, and Vertical Websites

Project Editor: Jennifer Tebbe

Acquisitions Editor: Erin Calligan Mooney

Senior Copy Editor: Danielle Voirol

Assistant Editor: David Lutton

Editorial Program Coordinator: Joe Niesen

Technical Editor: Chang Liu, Dan Moore

Editorial Manager: Christine Meloy Beck

Editorial Assistant: Alexa Koschier, Rachelle S. Amick

Cover Photo: © VOLODYMYR GRINKO/ iStockphoto.com

Composition Services

Project Coordinator: Katherine Crocker

Layout and Graphics: Carl Byers, Carrie A. Cesavice, Amy Hassos, Joyce Haughey, Christin Swinford

Proofreaders: Tricia Liebig, Dwight Ramsey

Indexer: Ty Koontz

Publishing and Editorial for Consumer Dummies

    Kathleen Nebenhaus, Vice President and Executive Publisher

    David Palmer, Associate Publisher

    Kristin Ferguson-Wagstaffe, Product Development Director

Publishing for Technology Dummies

    Andy Cummings, Vice President and Publisher

Composition Services

    Debbie Stailey, Director of Composition Services

Contents at a Glance

Introduction *1*

Part I: Getting Started with Circuit Analysis *5*

Chapter 1: Introducing Circuit Analysis 7

Chapter 2: Clarifying Basic Circuit Concepts and Diagrams 15

Chapter 3: Exploring Simple Circuits with Kirchhoff's Laws 25

Chapter 4: Simplifying Circuit Analysis with Source Transformation
and Division Techniques 41

Part II: Applying Analytical Methods
for Complex Circuits *65*

Chapter 5: Giving the Nod to Node-Voltage Analysis 67

Chapter 6: Getting in the Loop on Mesh Current Equations 83

Chapter 7: Solving One Problem at a Time Using Superposition 95

Chapter 8: Applying Thévenin's and Norton's Theorems 113

Part III: Understanding Circuits with Transistors
and Operational Amplifiers *131*

Chapter 9: Dependent Sources and the Transistors That Involve Them 133

Chapter 10: Letting Operational Amplifiers Do the Tough Math Fast 155

Part IV: Applying Time-Varying Signals
to First- and Second-Order Circuits *173*

Chapter 11: Making Waves with Funky Functions 175

Chapter 12: Spicing Up Circuit Analysis with Capacitors and Inductors 193

Chapter 13: Tackling First-Order Circuits 211

Chapter 14: Analyzing Second-Order Circuits 233

Part V: Advanced Techniques and
Applications in Circuit Analysis *253*

Chapter 15: Phasing in Phasors for Wave Functions 255

Chapter 16: Predicting Circuit Behavior with Laplace Transform Techniques 273

Chapter 17: Implementing Laplace Techniques for Circuit Analysis 295

Chapter 18: Focusing on the Frequency Responses 313

Part VI: The Part of Tens 335

Chapter 19: Ten Practical Applications for Circuits.................337
Chapter 20: Ten Technologies Affecting Circuits 341

Index 345

Table of Contents

Introduction *1*

About This Book 1
Conventions Used in This Book..1
What You're Not to Read 2
Foolish Assumptions 2
How This Book Is Organized 2
    Part I: Getting Started with Circuit Analysis.............................2
    Part II: Applying Analytical Methods for Complex Circuits.............3
    Part III: Understanding Circuits with Transistors
        and Operational Amplifiers 3
    Part IV: Applying Time-Varying Signals
        to First- and Second-Order Circuits 3
    Part V: Advanced Techniques and
        Applications in Circuit Analysis 3
    Part VI: The Part of Tens 3
Icons Used in This Book 4
Where to Go from Here...4

Part 1: Getting Started with Circuit Analysis *5*

Chapter 1: Introducing Circuit Analysis.........................7

Getting Started with Current and Voltage.......................................7
    Going with the flow with current 8
    Recognizing potential differences with voltage........................9
    Staying grounded with zero voltage....................................9
    Getting some direction with the passive sign convention 10
Beginning with the Basic Laws 11
Surveying the Analytical Methods for More-Complex Circuits 11
Introducing Transistors and Operational Amplifiers............................12
Dealing with Time-Varying Signals, Capacitors, and Inductors.............13
Avoiding Calculus with Advanced Techniques 13

Chapter 2: Clarifying Basic Circuit Concepts and Diagrams 15

Looking at Current-Voltage Relationships 15
    Absorbing energy with resistors 16
        Applying Ohm's law to resistors.................................16
        Calculating the power dissipated by resistors....................18
    Offering no resistance: Batteries and short circuits 18
        Batteries: Providing power independently.........................19
        Short circuits: No voltage, no power.............................19

Facing infinite resistance: Ideal current
sources and open circuits...20
All or nothing: Combining open and
short circuits with ideal switches...............................20
Mapping It All Out with Schematics..21
Going in circles with loops...22
Getting straight to the point with nodes.......................24

Chapter 3: Exploring Simple Circuits with Kirchhoff's Laws........25
Presenting Kirchhoff's Famous Circuit Laws................................25
Kirchhoff's voltage law (KVL): Conservation of energy.................26
Identifying voltage rises and drops 26
Forming a KVL equation...27
Kirchhoff's current law (KCL): Conservation of charge 29
Tracking incoming and outgoing current 29
Calculating KCL..30
Tackling Circuits with KVL, KCL, and Ohm's Law........................31
Getting batteries and resistors to work together 31
Starting with voltage..32
Bringing in current...32
Combining device equations with KVL 33
Summarizing the results..34
Sharing the same current in series circuits...............................34
Climbing the ladder with parallel circuits 36
Describing total resistance using conductance 37
Using a shortcut for two resistors in parallel..............38
Finding equivalent resistor combinations 38
Combining series and parallel resistors 40

**Chapter 4: Simplifying Circuit Analysis with Source
Transformation and Division Techniques 41**
Equivalent Circuits: Preparing for the Transformation..............42
Transforming Sources in Circuits...45
Converting to a parallel circuit with a current source..................45
Changing to a series circuit with a voltage source.....................47
Divvying It Up with the Voltage Divider 49
Getting a voltage divider equation for a series circuit................49
Figuring out voltages for a series circuit
with two or more resistors 51
Finding voltages when you have multiple current sources 52
Using the voltage divider technique repeatedly...........................55
Cutting to the Chase Using the Current Divider Technique 57
Getting a current divider equation for a parallel circuit 57
Figuring out currents for parallel circuits 59
Finding currents when you have multiple voltage sources 60
Using the current divider technique repeatedly 63

Part II: Applying Analytical Methods for Complex Circuits.......................... 65

Chapter 5: Giving the Nod to Node-Voltage Analysis 67
Getting Acquainted with Node Voltages and Reference Nodes............67
Testing the Waters with Node Voltage Analysis 69
    What goes in must come out: Starting with KCL at the nodes.......70
    Describing device currents in terms
        of node voltages with Ohm's law 70
    Putting a system of node voltage equations in matrix form 72
    Solving for unknown node voltages...73
Applying the NVA Technique...74
    Solving for unknown node voltageswith a current source............74
    Dealing with three or more node equations....................................76
Working with Voltage Sources in Node-Voltage Analysis 80

Chapter 6: Getting in the Loop on Mesh Current Equations 83
Windowpanes: Looking at Meshes and Mesh Currents............................83
Relating Device Currents to Mesh Currents..84
Generating the Mesh Current Equations 86
    Finding the KVL equations first..87
    Ohm's law: Putting device voltages in terms of mesh currents 87
    Substituting the device voltages into the KVL equations..............88
    Putting mesh current equations into matrix form..........................89
    Solving for unknown currents and voltages....................................89
Crunching Numbers: Using Meshes to Analyze Circuits 90
    Tackling two-mesh circuits...90
    Analyzing circuits with three or more meshes 92

Chapter 7: Solving One Problem at a Time Using Superposition 95
Discovering How Superposition Works..95
    Making sense of proportionality...96
    Applying superposition in circuits 98
    Adding the contributions of each independent source................100
Getting Rid of the Sources of Frustration..101
    Short circuit: Removing a voltage source......................................101
    Open circuit: Taking out a current source 102
Analyzing Circuits with Two Independent Sources 103
    Knowing what to do when the sources are two voltage sources 103
    Proceeding when the sources are two current sources 105
    Dealing with one voltage source and one current source............107
Solving a Circuit with Three Independent Sources.................................108

Chapter 8: Applying Thévenin's and Norton's Theorems **113**

Showing What You Can Do with Thévenin's
and Norton's Theorems 114
Finding the Norton and Thévenin Equivalents
for Complex Source Circuits..115
Applying Thévenin's theorem 117
Finding the Thévenin equivalent of a circuit
with a single independent voltage source........................117
Applying Norton's theorem 119
Using source transformation to find Thévenin or Norton............122
A shortcut: Finding Thévenin or Norton
equivalents with source transformation 122
Finding the Thévenin equivalent of a circuit
with multiple independent sources 122
Finding Thévenin or Norton with superposition 124
Gauging Maximum Power Transfer: A Practical
Application of Both Theorems..127

**Part III: Understanding Circuits with Transistors
and Operational Amplifiers 131**

**Chapter 9: Dependent Sources and
the Transistors That Involve Them** **133**

Understanding Linear Dependent Sources: Who Controls What 134
Classifying the types of dependent sources.................................134
Recognizing the relationship between dependent
and independent sources...136
Analyzing Circuits with Dependent Sources 136
Applying node-voltage analysis 137
Using source transformation..138
Using the Thévenin technique 140
Describing a JFET Transistor with a Dependent Source 142
Examining the Three Personalities of Bipolar Transistors 145
Making signals louder with the common emitter circuit.............146
Amplifying signals with a common base circuit 149
Isolating circuits with the common collector circuit...................151

**Chapter 10: Letting Operational Amplifiers
Do the Tough Math Fast** **155**

The Ins and Outs of Op-Amp Circuits 155
Discovering how to draw op amps...156
Looking at the ideal op amp and its transfer characteristics 157
Modeling an op amp with a dependent source.............................158
Examining the essential equations for
analyzing ideal op-amp circuits 159

Looking at Op-Amp Circuits 160
    Analyzing a noninverting op amp 160
    Following the leader with the voltage follower................. 162
    Turning things around with the inverting amplifier.......... 163
    Adding it all up with the summer 164
    What's the difference? Using the op-amp subtractor 166
Increasing the Complexity of What You Can Do with Op Amps 168
    Analyzing the instrumentation amplifier 168
    Implementing mathematical equations electronically........ 170
    Creating systems with op amps 171

Part IV: Applying Time-Varying Signals to First- and Second-Order Circuits 173

Chapter 11: Making Waves with Funky Functions 175

Spiking It Up with the Lean, Mean Impulse Function.................. 176
    Changing the strength of the impulse 178
    Delaying an impulse... 178
    Evaluating impulse functions with integrals 179
Stepping It Up with a Step Function 180
    Creating a time-shifted, weighted step function 181
    Being out of step with shifted step functions 182
    Building a ramp function with a step function.................. 182
Pushing the Limits with the Exponential Function 184
Seeing the Signs with Sinusoidal Functions 186
    Giving wavy functions a phase shift.................................. 187
    Expanding the function and finding Fourier coefficients............. 189
    Connecting sinusoidal functions to exponentials
        with Euler's formula... 190

Chapter 12: Spicing Up Circuit Analysis with Capacitors and Inductors 193

Storing Electrical Energy with Capacitors.................................. 193
    Describing a capacitor 194
    Charging a capacitor (credit cards not accepted) 195
    Relating the current and voltage of a capacitor 195
    Finding the power and energy of a capacitor.................... 196
    Calculating the total capacitance for
        parallel and series capacitors 199
        Finding the equivalent capacitance
            of parallel capacitors 199
        Finding the equivalent capacitance
            of capacitors in series.. 200
Storing Magnetic Energy with Inductors 200
    Describing an inductor.. 201
    Finding the energy storage of an attractive inductor 202

Calculating total inductance for series
and parallel inductors 203
Finding the equivalent inductance
for inductors in series.......................................203
Finding the equivalent inductance
for inductors in parallel....................................204
Calculus: Putting a Cap on Op-Amp Circuits.............................205
Creating an op-amp integrator.....................................205
Deriving an op-amp differentiator 207
Using Op Amps to Solve Differential Equations Really Fast..................208

Chapter 13: Tackling First-Order Circuits 211

Solving First-Order Circuits with Diff EQ....................................211
Guessing at the solution with the
natural exponential function 213
Using the characteristic equation for a first-order equation 214
Analyzing a Series Circuit with a Single Resistor and Capacitor...........215
Starting with the simple RC series circuit 215
Finding the zero-input response 217
Finding the zero-state response by
focusing on the input source....................................219
Adding the zero-input and zero-state responses
to find the total response...222
Analyzing a Parallel Circuit with a Single Resistor and Inductor 224
Starting with the simple RL parallel circuit.....................................225
Calculating the zero-input response for an RL parallel circuit 226
Calculating the zero-state response for an RL parallel circuit.....228
Adding the zero-input and zero-state responses
to find the total response...230

Chapter 14: Analyzing Second-Order Circuits 233

Examining Second-Order Differential Equations
with Constant Coefficients...233
Guessing at the elementary solutions:
The natural exponential function..............................235
From calculus to algebra: Using the characteristic equation 236
Analyzing an RLC Series Circuit..236
Setting up a typical RLC series circuit 237
Determining the zero-input response 239
Calculating the zero-state response.......................................242
Finishing up with the total response...245
Analyzing an RLC Parallel Circuit Using Duality.....................................246
Setting up a typical RLC parallel circuit...........................247
Finding the zero-input response 249
Arriving at the zero-state response 250
Getting the total response 251

Part V: Advanced Techniques and Applications in Circuit Analysis................................. 253

Chapter 15: Phasing in Phasors for Wave Functions.............255
Taking a More Imaginative Turn with Phasors...256
   Finding phasor forms 256
   Examining the properties of phasors..458
Using Impedance to Expand Ohm's
Law to Capacitors and Inductors..259
   Understanding impedance...260
   Looking at phasor diagrams 261
   Putting Ohm's law for capacitors in phasor form 262
   Putting Ohm's law for inductors in phasor form..........................263
Tackling Circuits with Phasors..263
   Using divider techniques in phasor form...................................264
   Adding phasor outputs with superposition 266
   Simplifying phasor analysis with Thévenin and Norton..............268
   Getting the nod for nodal analysis..270
   Using mesh-current analysis with phasors 271

Chapter 16: Predicting Circuit Behavior with Laplace Transform Techniques 273
Getting Acquainted with the Laplace Transform
   and Key Transform Pairs 273
Getting Your Time Back with the Inverse Laplace Transform...............276
   Rewriting the transform with partial fraction expansion 276
   Expanding Laplace transforms with complex poles 278
   Dealing with transforms with multiple poles 280
Understanding Poles and Zeros of F(s) 282
Predicting the Circuit Response with Laplace Methods 285
   Working out a first-order RC circuit 286
   Working out a first-order RL circuit 290
   Working out an RLC circuit 292

Chapter 17: Implementing Laplace Techniques for Circuit Analysis 295
Starting Easy with Basic Constraints 296
   Connection constraints in the s-domain...................................296
   Device constraints in the s-domain 297
      Independent and dependent sources...................................297
      Passive elements: Resistors, capacitors, and inductors.....297
      Op-amp devices...299
   Impedance and admittance 299
Seeing How Basic Circuit Analysis Works in the s-Domain...................300
   Applying voltage division with series circuits 300
   Turning to current division for parallel circuits...........................302

Conducting Complex Circuit Analysis in the s-Domain 303

Using node-voltage analysis 303

Using mesh-current analysis 304

Using superposition and proportionality 305

Using the Thévenin and Norton equivalents 309

Chapter 18: Focusing on the Frequency Responses 313

Describing the Frequency Response and Classy Filters 314

Low-pass filter 315

High-pass filter 316

Band-pass filters 316

Band-reject filters 317

Plotting Something: Showing Frequency Response à la Bode 318

Looking at a basic Bode plot 319

Poles, zeros, and scale factors: Picturing
Bode plots from transfer functions 320

Turning the Corner: Making Low-Pass and
High-Pass Filters with RC Circuits 325

First-order RC low-pass filter (LPF) 325

First-order RC high-pass filter (HPF) 326

Creating Band-Pass and Band-Reject Filters
with RLC or RC Circuits 327

Getting serious with RLC series circuits 327

RLC series band-pass filter (BPF) 327

RLC series band-reject filter (BRF) 330

Climbing the ladder with RLC parallel circuits 330

RC only: Getting a pass with a band-pass
and band-reject filter 332

Part VI: The Part of Tens 335

Chapter 19: Ten Practical Applications for Circuits 337

Potentiometers 337

Homemade Capacitors: Leyden Jars 338

Digital-to-Analog Conversion Using Op Amps 338

Two-Speaker Systems 338

Interface Techniques Using Resistors 338

Interface Techniques Using Op Amps 339

The Wheatstone Bridge 339

Accelerometers 339

Electronic Stud Finders 340

555 Timer Circuits 340

Chapter 20: Ten Technologies Affecting Circuits. 341

Smartphone Touchscreens 341
Nanotechnology...341
Carbon Nanotubes...342
Microelectromechanical Systems 342
Supercapacitors...343
The Memristor 343
Superconducting Digital Electronics..343
Wide Bandgap Semiconductors...343
Flexible Electronics 344
Microelectronic Chips that Pair Up with Biological Cells 344

Index **345**

Introduction

• •

Circuit analysis is often one of those weed-out classes in engineering schools. Either you pass the class to study engineering, or you don't pass and start thinking about something else. Well, I don't want you to get weeded out, because engineering is such a rewarding field. This book is here to help you make sense of circuit analysis concepts that may be puzzling you. Along the way, you explore a number of analytical tools that give you short-cuts and insight into circuit behavior.

You can take the tools you find here and apply them to whatever high-tech gizmo or craze is out there. And not only can you pass your class, but you can also take these concepts to the real world, enriching human lives with comfort and convenience and rewarding you with more time to do useful activities.

About This Book

Like all other *For Dummies* books, *Circuit Analysis For Dummies* isn't a tutorial. Rather, it's a reference book, which means you don't have to read it from cover to cover, although you certainly can if that's your preference. You can jump right to the topics or concepts you're having trouble with. Either way, you'll find helpful information along with some real-world examples of electrical concepts that may be hard to visualize otherwise.

Conventions Used in This Book

I use the following conventions throughout the text to make things consistent and easy to understand:

- ✔ New terms appear in *italics* and are closely followed by an easy-to-understand definition. Variables likewise appear in italics.

- ✔ **Bold** is used to highlight keywords in bulleted lists and the action parts of numbered steps. It also indicates vectors.

- ✔ Lowercase variables indicate signals that change with time, and upper-case variables indicate signals that are constant. For example, $v(t)$ and $i(t)$ denote voltage and current signals that change with time. If, however V and I are capitalized, then those signals don't vary in time.

What You're Not to Read

Although it'd be great if you read every word, you're welcome to skip the sidebars (the shaded boxes sprinkled throughout the book) and paragraphs flagged with a Technical Stuff icon.

Foolish Assumptions

I may be going out on a limb, but as I wrote this book, here's what I assumed about you:

- ✔ You're currently taking an introductory circuit analysis course, and you need help with certain concepts and techniques. Or you're planning to take a circuit analysis course in the next semester, and you want to be prepared with some supplementary material.
- ✔ You have a good grasp of linear algebra and differential equations.
- ✔ You've taken an introductory physics class, which exposed you to the concepts of power, positive and negative charges, voltage, and current.

How This Book Is Organized

Circuit analysis integrates a variety of topics from your math and physics courses, and it introduces a variety of techniques to solve for circuit behavior. To help you grasp the concepts in manageable bites, I've split the book into several parts, each consisting of chapters on related topics.

Part 1: Getting Started with Circuit Analysis

This part gives you the engineering lingo, concepts, and techniques necessary for tackling circuit analysis. Here, I help you quickly grasp the main aspects of circuit analysis so you can analyze circuits, build things, and predict what's going to happen. If you're familiar with current, voltage, power, and Ohm's and Kirchhoff's laws, you can use this part as a refresher.

Part II: Applying Analytical Methods for Complex Circuits

This part looks at general analytical methods to use when dealing with more complicated circuits. When you have many simultaneous equations to solve or too many inputs, you can use various techniques to reduce the number of equations and simplify circuits to a manageable level.

Part III: Understanding Circuits with Transistors and Operational Amplifiers

This part deals with two devices that require power to make them work. You can use transistors as current amplifiers, and you can use operational amplifiers as voltage amplifiers.

Part IV: Applying Time-Varying Signals to First- and Second-Order Circuits

This part gets tougher because you're dealing with changing signals and with circuits that have passive energy-storage devices such as inductors and capacitors. You also need to know differential equations in order to analyze circuit behavior for first- and second-order circuits.

Part V: Advanced Techniques and Applications in Circuit Analysis

This part takes the problems described in Part IV and changes a calculus-based problem into one requiring only algebra. You do this conversion by using phasor and Laplace techniques. You can gather additional insight into circuit behavior from the poles and zeros of an equation, which shape the frequency response of circuits called filters.

Part VI: The Part of Tens

Here you find out about ten applications and ten technologies that make circuits more interesting.

Icons Used in This Book

To make this book easier to read and simpler to use, I include some icons to help you find key information.

Anytime you see this icon, you know the information that follows will be worth recalling after you close this book — even if you don't remember anything else you just read.

This icon appears next to information that's interesting but not essential. Don't be afraid to skip these paragraphs.

This bull's-eye points out advice that can save you time when analyzing circuits.

This icon is here to prevent you from making fatal mistakes in your analysis.

Where to Go from Here

This book isn't a novel — you can start at the beginning and read it through to the end, or you can jump right in the middle. If you like the calculus approach to solving circuits, head to the chapters on first- and second-order circuits. If calculus doesn't suit your fancy or if you're itching to find out what the Laplace transform is all about, flip straight to Chapter 16.

If you're not sure where to start, or you don't know enough about circuit analysis to even *have* a starting point in mind yet, no problem — that's exactly what this book is for. Just hop right in and get your feet wet. I recommend starting with the chapters in Part I and moving forward from there.

Part I
Getting Started with Circuit Analysis

In this part . . .

- ✔ Discover what circuit analysis is all about.

- ✔ Get the scoop on current and voltage behaviors in common circuit components and find out how to read circuit diagrams.

- ✔ Familiarize yourself with Kirchhoff's voltage law and Kirchhoff's current law — two laws essential for creating connection equations.

- ✔ Use source transformation and current and voltage divider techniques to simplify circuit analysis.

Chapter 1

Introducing Circuit Analysis

In This Chapter
▶ Understanding current and voltage
▶ Applying laws when you connect circuit devices
▶ Analyzing circuits with algebra and calculus
▶ Taking some mathematical shortcuts

Circuit analysis is like the psychoanalysis of the electrical engineering world because it's all about studying the behavior of circuits. With any circuit, you have an input signal, such as a battery source or an audio signal. What you want to figure out is the circuit's *output* — how the circuit responds to a given input.

A circuit's output is either a voltage or a current. You have to analyze the voltages and currents traveling through each element or component in the circuit in order to determine the output, although many times you don't have to find *every* voltage and *every* current within the circuit.

Circuit analysis is challenging because it integrates a variety of topics from your math and physics courses in addition to introducing techniques specific to determining circuit behavior. This chapter gives you an overview of circuit analysis and some of the key concepts you need to know before you can begin understanding circuits.

Getting Started with Current and Voltage

Being able to analyze circuits requires having a solid understanding of how voltage and current interact within a circuit. Chapter 2 gives you insight into how voltage and current behave in the types of devices normally found in circuits, such as resistors and batteries. That chapter also presents the basic features of circuit diagrams, or *schematics*.

The following sections introduce you to current and voltage as well as a direction-based convention that's guaranteed to come in handy in circuit analysis.

Going with the flow with current

Current is a way of measuring the amount of electric charge passing through a given point within a certain amount of time. Current is a flow rate. The mathematical definition of a current is as follows:

$$i = \frac{dq}{dt}$$

The variable i stands for the current, q stands for the electrical charge, and t stands for time.

The charge of one electron is 1.609×10^{-19} coulombs (C).

Current measures the flow of charges with dimensions of coulombs per second (C/s), or *amperes* (A). In engineering, the current direction describes the net flow of positive charges. Think of current as a *through variable,* because the flow of electrical charge passes through a point in the circuit. The arrow in Figure 1-1 shows the current direction.

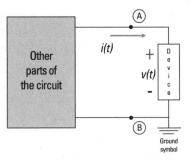

Figure 1-1:
Current
direction,
voltage
polarities,
and the
passive sign
convention.

Illustration by Wiley, Composition Services Graphics

Measuring current through a device requires just one point of measurement. As an analogy, say you're asked to count the number of cars flowing through your long stretch of residential street for 10 minutes. You can count the number of cars from your home or your friend's home next door or the house across the street. You need just one location point to measure the flow of cars.

Two types of current exist: alternating current (AC) and direct current (DC). With AC, the charges flow in both directions. With DC, the charges flow in just one direction.

If you have trouble keeping AC and DC straight, try this mnemonic device: AC means "always changing," and DC means "doesn't change."

Recognizing potential differences with voltage

From physics, you know that plus and minus charges attract each other and that like charges repel each other. You need energy to separate the opposite charges. As long as the charges are separated, they have electric potential energy.

Voltage measures the amount of energy w required to move a given amount of charge q as it passes through the circuit. You can think of voltage as electric potential difference. Mathematically, voltage is defined as

$$v = \frac{dw}{dq}$$

Voltage has units of *volts* (V), which is the same as joules per coulomb (J/C). In a 12-volt car battery, the opposite charges on the battery terminals have a separation of 12 units of energy per unit of charge (or 12 V = 12 J/C). When terminals are separated, there's no current flow. If you provide a conducting path between the opposite charges, you now have charges flowing, resulting in an electric current.

It takes two points to measure voltage across a device, just like it takes two points to measure height or distance. That's why you can think of voltage as an *across variable*.

Refer to Figure 1-1 to see the positive and negative voltage signs (called *polarities*) of a device, labeled at Terminals A and B.

Staying grounded with zero voltage

Because measuring voltage requires two points, you need a common reference point called a *ground*. You assign ground as 0 volts, where all other points in a circuit are measured with respect to ground. This is analogous to defining sea level as a reference point of 0 feet so you can measure the height of mountains. When the sea canyon is below sea level, you assign a negative algebraic sign. In circuits, the negative sign means the answer is less than the ground potential of 0 volts. Refer to Figure 1-1 to see an example of a typical ground symbol.

From algebra to calculus and back to algebra

When you're looking at simple circuits, such as a direct current (DC) circuit that involves only resistors and a constant battery source, you can get by with just algebra in your analysis. Because you don't have to worry about any fancy math, you can focus on analytical approaches such as node-voltage analysis (Chapter 5) and superposition (Chapter 7).

But when you start looking at more complex circuits, such as an alternating current (AC) circuit, the math becomes more complex. AC circuits have time-varying sources, capacitors, and inductors, so you need calculus to

deal with the changes of electrical variables over time. Applying connection constraints (Kirchhoff's laws) to AC circuits gives you differential equations, but never fear! The chapters in Part IV show you how to solve differential equations of first-order and second-order circuits when you have capacitors and inductors connected to resistors.

Of course, sometimes you can use advanced techniques to skip over calculus entirely. You can convert differential equations into simpler algebraic problems using the Laplace transform method, which I introduce in Chapter 16.

 You may have seen a streetcar with one trolley pole and an electric bus with two trolley poles. Why the difference? The streetcar uses its wheels as its ground point, so current coming from the generator flows back to the generator through the earth via the wheels grounded at 0 volts. For the electric bus, you need two wires because current can't flow through the rubber tires.

Getting some direction with the passive sign convention

Along with the algebraic signs of your calculated answers, the *passive sign convention* orients you to what's happening in a circuit. Specifically, it tells you that current enters a passive device at its positive voltage terminal.

Here's how passive sign convention works: You assign plus and minus signs to each device to serve as reference marks. After you arbitrarily assign the polarities of a device, you define the current direction so that it enters the positive side. (You can see what I mean by referring to Figure 1-1.) If your answers for voltage or current are positive, then the polarities line up with your assigned polarities or current direction. If your answers come up negative, they're opposite to your assigned polarities or current direction. A negative answer isn't wrong; it's just reverse to your assigned reference marks.

 The way you assign your polarities and current direction doesn't control the circuit behavior. Rather, the algebraic signs of your answers tell you the actual directions of voltages and current in the circuit.

Beginning with the Basic Laws

A circuit is basically a collection of electrical devices, such as resistors, batteries, capacitors, and inductors, arranged to perform a certain function. Each component of a circuit has its own constraints. When you connect devices in any circuit, the devices follow certain laws:

- ✔ **Ohm's law:** This law describes a linear relationship between the voltage and current for a resistor. You can find details about resistors and Ohm's law in Chapter 2.

- ✔ **Kirchhoff's voltage law (KVL):** KVL says the algebraic sum of the voltage drops and rises around a loop of a circuit is equal to zero. You can find an explanation of voltage drops, voltage rises, circuit loops, and KVL in Chapter 3.

- ✔ **Kirchhoff's current law (KCL):** KCL says the algebraic sum of incoming and outgoing currents at a node is equal to zero. Chapter 3 provides info on applying KCL and defines nodes in a circuit.

With these three laws, you can solve for the current or voltage in any device.

Applying Kirchhoff's laws can become tedious, but you can take some shortcuts. Source transformation allows you to convert circuits to either parallel or series circuits. Then, with all the devices connected in series or in parallel, you can use the voltage divider and current divider techniques to find the voltage or current for any device. I cover these techniques in Chapter 4.

Surveying the Analytical Methods for More-Complex Circuits

When you have many simultaneous equations to solve or too many inputs, you can use the following techniques to reduce the number of simultaneous equations and simplify the analysis:

- ✔ **Node-voltage analysis:** A node is a point in the circuit. This technique has you apply Kirchhoff's current law (KCL), producing a set of equations that you use to find unknown node voltages. When you know all the node voltages in a circuit, you can find the voltage across each device. I cover node-voltage analysis in Chapter 5.

✔ **Mesh-current analysis:** Mesh-current analysis deals with circuits that have many devices connected in many loops. You use Kirchhoff's voltage law (KVL) to develop a set of equations with unknown mesh currents. Because you can describe the device currents in terms of the mesh currents, finding the mesh currents lets you calculate the current through each device in the circuit. See Chapter 6 for info on mesh-current analysis.

✔ **Superposition:** When you have multiple independent power sources in a linear circuit, superposition comes to your rescue. Analyzing linear circuits involves using only devices (such as resistors, capacitors, and inductors) and independent sources. By applying superposition, you can take a complex circuit that has multiple independent sources and break it into simpler circuits, each with only one independent source. The circuit's total output then is the algebraic sum of output contributions due to the input from each independent source. Turn to Chapter 7 for details on superposition.

✔ **Thévenin's and Norton's theorems:** Thévenin and Norton equivalent circuits are valuable tools when you're connecting and analyzing two parts of a circuit. The interaction between the source circuit (which processes and delivers a signal) and load circuits (which consume the delivered signal) offers a major challenge in circuit analysis. Thévenin's theorem simplifies the analysis by replacing the source circuit's complicated arrangement of independent sources and resistors with a single voltage source connected in series with a single resistor. Norton's theorem replaces the source circuit with a single current source connected in parallel with a single resistor. You can find out more about both theorems, including how to apply them, in Chapter 8.

Introducing Transistors and Operational Amplifiers

Although transistors and operational amplifiers (op amps) are modeled with dependent sources, they're referred to as *active devices* because they require power to work. Transistors, which are made of semiconductor material, are used primarily as current amplifiers (see Chapter 9). Op amps are linear devices consisting of many transistors, resistors, and capacitors. They're used to perform many mathematical and processing operations, including voltage amplification (see Chapter 10). You can think of op amps as very high-gain DC amplifiers.

The op amp is one of the leading linear active devices in modern circuit applications. This device does mathematical operations (addition, subtraction, multiplication, division, integration, derivatives, and so on) quickly because it does them electronically. You put together basic op-amp circuits to build mathematical models.

Dealing with Time-Varying Signals, Capacitors, and Inductors

Circuits deal with signals that carry energy and information. *Signals* are time-varying electrical quantities processed by the circuit. Throughout the book, you deal with linear circuits, where the output signal is proportional to the input signal.

Chapter 11 introduces you to signal sources that change with time (unlike batteries, whose signals don't change with time). Signals that change in time can carry information about the real world, like temperature, pressure, and sound. You can combine basic functions such as sine and exponential functions to create even more interesting signals.

When you add passive, energy-storing elements (such as capacitors and inductors) to a circuit, the analysis gets a little tougher because now you need differential equations to analyze the circuit's behavior. In fact, the circuits created with capacitors and inductors get their names from the differential equations that result when you apply Kirchhoff's laws in the course of analysis:

✔ *First-order circuits,* which have a resistor and capacitor or a resistor and inductor, are described with first-order differential equations. The capacitor's current is related to the first derivative of the voltage across the capacitor, and the inductor's voltage is related to the first derivative of the current through the inductor. See Chapter 13 for help analyzing first-order circuits.

✔ *Second-order circuits* consist of capacitors, inductors, and resistors and are described by second-order differential equations. Flip to Chapter 14 for pointers on analyzing these circuits.

Avoiding Calculus with Advanced Techniques

I don't know about you, but I hate using calculus when I don't have to, which is why I'm a fan of the advanced circuit analysis techniques that allow you to convert calculus-based problems into problems requiring only algebra.

Phasors make your life simple when you're dealing with circuits that have capacitors and inductors, because you don't need differential equations to analyze circuits in the phasor domain. Phasor analysis investigates circuits that have capacitors and inductors in the same way you analyze circuits that have only resistors. This technique applies when your input is a sine wave (or a sinusoidal signal). See Chapter 15 for details on phasors.

Analyzing circuits with software

When circuits get too complex to analyze by hand, today's software offers many capabilities. Here are some commonly used software tools:

✔ **SPICE:** This software was originally developed at the Electronics Research Laboratory of the University of California, Berkeley. SPICE stands for Simulation Program for Integrated Circuit Emphasis. PSpice is a PC version of SPICE, and several companies, such as Cadence (www.cadence.com) and Linear Technology (www.linear.com), produce various versions of SPICE. Both companies offer demos and free versions of their software.

✔ **National Instrument's Multisim:** This is one of the granddaddies of circuit analysis software as well as a great tool for beginners. It also has a cool feature that shows changing voltages in real time of the circuit. The trial version pretty much lets you do anything you want. A student version is available as well. Visit www.ni.com/multisim/ for more information.

✔ **Ngspice:** This tool is a widely used open-source circuit simulator from Sourceforge. The free Ngspice software (available at ngspice.sourceforge.net) is developed by many users, and its code is based on several major open-source software packages.

Chapter 16 describes a more general technique that's handy when your input isn't a sinusoidal signal: the Laplace transform technique. You use the Laplace transform to change a tough differential equation into a simpler problem involving algebra in the Laplace domain (or s-domain). You can then study the circuit's behavior using only algebra. The s-domain method I cover in Chapter 17 gives you the same results you'd get from calculus methods to solve differential equations, which you find in Chapters 13 and 14. The algebraic approach in the s-domain follows along the same lines as the approach you use for resistor-only circuits, only in place of resistors, you have s-domain impedances.

A major component found in older entertainment systems is an electronic filter that shapes the frequency content of signals. In Chapter 18, I present low-pass, high-pass, band-pass, and band-stop (or band-reject) filters based on simple circuits. This serves as a foundation for more-complex filters to meet more stringent requirements.

Chapter 18 also covers Bode diagrams to describe the frequency response of circuits. The Bode diagrams help you visualize how poles and zeros affect the frequency response of a circuit. The frequency response is described by a transfer (or network) function, which is the ratio of the output signal to the input signal in the s-domain. The *poles* are the roots of the polynomial in the transfer function's denominator, and the *zeros* are the roots of the polynomial in the numerator.

Chapter 2

Clarifying Basic Circuit Concepts and Diagrams

In This Chapter

▶ Sorting out current-voltage relationships

▶ Mapping out circuits with schematics

▶ Understanding a circuit's loops and nodes

*B*efore you can begin working with circuits, you need to have a basic understanding of how current and voltage behave in some of the devices most commonly found in circuits. You also need to be able to read basic circuit diagrams, or *schematics.* This chapter is all about helping you get comfortable with these basics so you can dive confidently into the world of circuit analysis.

Looking at Current-Voltage Relationships

Given that power is a rate of energy transfer, electrical power $p(t)$ is defined as the product of the voltage $v(t)$ and current $i(t)$ as a function of time:

$$p(t) = i(t) \cdot v(t)$$

To remember the formula $p = iv$, I tell students to remember the phrase *poison ivy.* It may be corny, but it works.

An electrical device absorbs power when $p(t)$ is positive, implying that the current and voltage have the same algebraic sign. The device delivers power when $p(t)$ is negative, implying current and voltage have opposite algebraic signs. See Chapter 1 for details on the passive sign convention and what negative current and negative voltage mean.

Power has units of *watts* (W), or joules per second. The units of current (coulombs per second) and voltage (joules per coulomb) should cancel out to give you the desired units for power. Here's the dimensional analysis:

$$p(t) = i(t) \cdot v(t) \quad \rightarrow \quad \left(\frac{\text{joules}}{\text{sec}} \right) = \left(\frac{\cancel{\text{coulombs}}}{\text{sec}} \right) \cdot \left(\frac{\text{joules}}{\cancel{\text{coulombs}}} \right)$$

So the power relationship works out as far as units are concerned!

Because power involves current and voltage, understanding the current-voltage (*i-v*) characteristics of various devices, such as resistors and batteries, is important. Resistors have a very straightforward relationship with voltage and current. In fact, for circuits that contain only resistors and independent power sources, the relationship between current and voltage simply depends on a device's resistance, which is a constant *R*. In the following sections, I introduce you to some devices and circuit configurations that provide a certain amount of resistance, no resistance, or infinite resistance.

Absorbing energy with resistors

Resistors are simple electrical devices that appear in almost every circuit. They suck up energy and give it off as heat. An everyday object like a toaster or an incandescent light bulb can be modeled as a resistor.

You may think resistors don't do much because they waste energy, but they actually have a few important purposes:

- **Reducing voltage:** A resistor can use up some voltage so that not all of the supplied voltage falls on another device. You're basically dividing the supplied voltage into smaller voltages by adding resistors to a circuit.

- **Limiting current:** If you don't want current to burn up a device, you can limit current by connecting a resistor to the device.

- **Timing and filtering:** You can use resistors, along with capacitors, to create timing circuits or filters. I discuss timing and filtering in Chapters 12 and 13 and filtering specifically in Chapter 18.

The following sections introduce current-voltage relationships and graph the *i-v* curves for resistors. They also show you how to calculate the power dissipated as heat.

Applying Ohm's law to resistors

Ohm's law says that the current through a linear resistor is proportional to the voltage across the resistor. Mathematically, you have Ohm's law described as

$$v = Ri$$

$$i = Gv \qquad \rightarrow \qquad R = \frac{1}{G}$$

where v is voltage, i is current, R is resistance, and G is conductance. The resistance R or conductance G is a proportionality constant relating the resistor voltage and its current. For example, if the voltage is doubled, then the current is doubled.

Resistance provides a measure of difficulty in pushing electricity through a circuit. The unit of resistance is *ohms* (Ω), and the unit of conductance is *siemens* (S). For fun (if you call algebra fun), you can rearrange Ohm's law:

$$i = \frac{v}{R}$$

$$R = \frac{v}{i}$$

When you don't know the current, use the top equation, and when you don't know the resistance, use the bottom equation.

Figure 2-1 shows the symbol and *i-v* characteristic for a linear resistor. The slope of the line gives you conductance G, and the reciprocal of the slope produces the resistance value R.

You can have large current flow for a small applied voltage if the resistance is small enough. Some materials cooled to very low temperatures are superconductors, having near-zero resistance. As soon as current flows in a superconducting circuit, current flows forever unless you disconnect the voltage source.

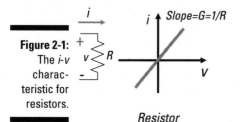

Figure 2-1:
The *i-v* characteristic for resistors.

Illustration by Wiley, Composition Services Graphics

Calculating the power dissipated by resistors

Because power is $p = iv$, you can use Ohm's law, $v = iR$, to figure the amount of heat a resistor gives off when current flows or voltage is applied across the resistor. Here are two versions of the power-dissipation formula, which you get by plugging in the voltage or current value from Ohm's law into $p = iv$:

$$p = iv = i(iR) = i^2 R$$

$$p = \left(\frac{v}{R}\right)v = \frac{v^2}{R}$$

So by knowing either the voltage or current for a given resistor R, you can find the amount of power dissipated. If you calculate the power dissipated as 0.1 watts, then a ¼-watt (0.25-watt) resistor can handle this amount of power. A ⅛-watt (0.125-watt) resistor should be able to handle that amount as well, but when it comes to power ratings, err on the larger side.

Offering no resistance: Batteries and short circuits

Batteries and short circuits have different i-v characteristics but the same slope (or equivalently, zero resistance), as Figure 2-2 shows. In certain situations, you can remove a battery from a circuit by replacing it with a short circuit, 0 volts (I explain how in Chapter 7). Read on for the details on batteries, short circuits, and their i-v characteristics.

Figure 2-2:
You get zero resistance and constant voltage from an ideal voltage source or short circuit.

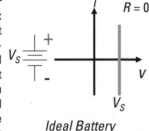

Ideal Battery (Voltage Source)

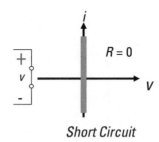

Short Circuit

Illustration by Wiley, Composition Services Graphics

Short stuff: Why birds on a wire don't get zapped

Have you ever wondered how birds sit on a bare high-voltage line of 25,000 volts without getting shocked? The entire length of wire is at 25,000 volts, so the entire bird on the wire is also at 25,000 volts. Because there's no voltage difference, current doesn't flow through the bird. Now, if the bird decides to stretch its wing and touches an adjacent wire at 0 volts, well, bad things happen, and it's bye-bye birdie! Fortunately for the birds, the power lines are strung apart so they aren't short circuited.

Batteries: Providing power independently

In circuit analysis, batteries are referred to as *independent sources.* Specifically, batteries are *independent sources of voltage,* supplying the circuit with a constant voltage that's independent of the current. So no matter how much current is drawn from the battery, you still have the same voltage. Figure 2-2 shows the electrical symbol and the *i-v* characteristic of a battery. Because the slope is infinite, an ideal battery has zero resistance.

You can convert a battery into an independent current source through source transformation, as I explain in Chapter 4. I cover independent current sources later in "Facing infinite resistance: Ideal current sources and open circuits."

Short circuits: No voltage, no power

Figure 2-2 shows that, like a battery, a short circuit has an infinite slope (and therefore infinite resistance) in its *i-v* characteristic. And just like a battery, the voltage is also constant: In a short circuit, there's zero voltage across a wire, no matter how much current flows through it. Because there's no voltage across a short circuit, there's zero absorbed power ($p = 0$ watts).

When you connect two points in a circuit that have different voltages, you get a *short circuit.* When this happens, you bypass the other parts of a circuit (called the *load*) and establish a path of low resistance, causing most of the current to flow around or away from some other parts in the circuit. Accidental short circuits, especially between the high and low voltages of a power supply, can cause strong current to flow, possibly damaging or overheating the power supply and the circuit if the circuit isn't protected by a fuse.

Facing infinite resistance: Ideal current sources and open circuits

Figure 2-3 shows that an ideal current source and an open circuit both have zero slope in their *i-v* characteristics, meaning that they have infinite resistance. And in both cases, the current is constant.

The infinite resistance makes sense because if current entered the ideal current source, the current would no longer be constant. In circuit analysis, you can remove a current source by replacing it with an infinite resistor or open circuit. (You can read more about this change in Chapter 7.)

Figure 2-3:
You get infinite resistance and constant current from an ideal current source or open circuit.

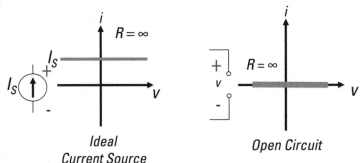

Ideal
Current Source

Open Circuit

Illustration by Wiley, Composition Services Graphics

An *open circuit* occurs when there's no current flow for any applied voltage, like when you blow a fuse. Because there's no current flow, there's no power absorbed (p = 0 watts) in an open-circuit device.

All or nothing: Combining open and short circuits with ideal switches

Think of ideal switches as a combination of an open circuit and a short circuit. When a switch is on, you have a short circuit, providing current flow in the circuit. When a switch is off, you have an open circuit, leaving zero current flow. Figure 2-4 illustrates an ideal switch's *i-v* characteristic along with its symbol, which shows the switch in the *off* state.

Because the switch has zero voltage in its *on* state and zero current in its *off* state, no power (p = 0 watts) is dissipated in the switch.

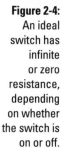
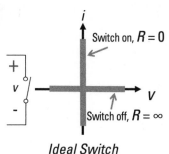

Figure 2-4:
An ideal
switch has
infinite
or zero
resistance,
depending
on whether
the switch is
on or off.

Illustration by Wiley, Composition Services Graphics

Mapping It All Out with Schematics

Schematics, which are drawings that symbolize a circuit, help you see the connections between electronic components. They also help you trouble-shoot your circuit design during construction. You usually arrange electronic schematics from top to bottom and left to right, following the path to place the components.

Schematics use symbols to represent the different components of circuits. Here are some basic symbols to help you get started:

✔ **Wires:** Simple conductors, or wires, appear as plain lines in schematics. When two wires cross each other, you know the following:

- If a dot appears at their intersection, the wires are connected (see the top-left diagram in Figure 2-5).

- If the dot is absent or you see a curved bridge over one of the wires, the wires are unconnected (see the top-right diagram in Figure 2-5).

Wires that cross over are found in more-complicated circuits, which appear much later in the book.

Lines don't necessarily depict actual wires, like the rat's nest of wires you'd see inside an old radio; the lines simply represent a pathway of conductors. Today you have the more common metallic pathway called *traces* on a board. If you've ever opened up a desktop computer, you've seen traces on a big motherboard and wires connecting various devices like power supplies, sound cards, and hard drives.

✔ **Gates:** Control lines at the gate terminals of switches are represented by dashed lines (see the bottom-left diagram in Figure 2-5). By applying a voltage to the gate terminal, you can control the *on* and *off* states of the switch.

✔ **Power supplies:** Power supplies in schematics incorporate the device symbols I show you in Figures 2-2 and 2-3. You see power supply connections at the bottom right of Figure 2-5. The left diagram shows a way to reduce the clutter found in schematics by not drawing the symbol for the power supply. The schematic on the right shows the ground symbol, which marks a reference point of 0 volts.

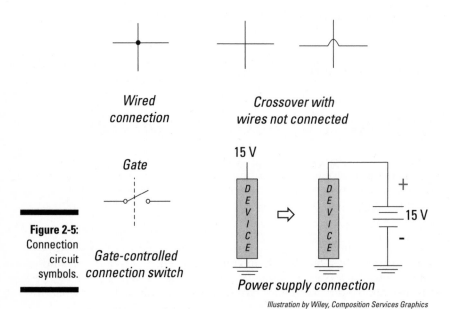

Figure 2-5:
Connection
circuit
symbols.

Wired connection

Crossover with wires not connected

Gate

Gate-controlled connection switch

Power supply connection

Illustration by Wiley, Composition Services Graphics

Additionally, circuit schematics often depict circular arrangements of electronic devices and junction points. The circular arrangements of electrical devices are called *loops,* and the junction or connection points are called *nodes.* I discuss these features next.

Going in circles with loops

When looking at a circuit schematic such as the one in Figure 2-6, you often see a collection of resistors and a battery connected together in some configuration. The loops form circular connections of devices. By definition, a *loop* occurs when you trace a closed path through the circuit in an orderly way, passing through each device only once.

This method of generating a closed path allows you to get consistent results when analyzing circuits. To form a loop or closed path, you must start at one point in the circuit and end up at the same place, much like going around the block in your neighborhood.

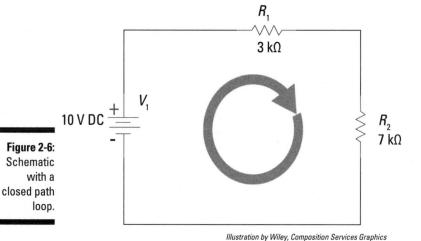

Figure 2-6:
Schematic
with a
closed path
loop.

Illustration by Wiley, Composition Services Graphics

As more devices are connected to the circuit, there's an increased likelihood that more loops will occur. Figure 2-7 shows a circuit with two inner loops (Loops 1 and 2) and one big outer loop (Loop 3).

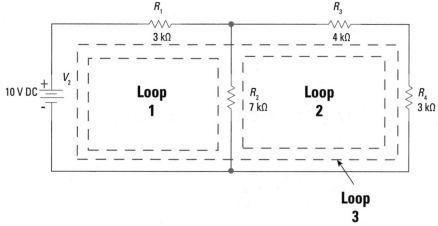

Figure 2-7:
Schematic
with three
loops.

Illustration by Wiley, Composition Services Graphics

Getting straight to the point with nodes

A *node* is simply a junction or point where two or more devices are connected. Be sure to add the following important points about nodes to your memory bank:

- ✔ A node isn't confined to a point; it includes the wire between devices.
- ✔ Wires connected to a node have zero resistance.

Figure 2-8, which depicts three nodes (or junctions), emphasizes the preceding points about nodes. The connected devices, which can be either resistors or independent sources (like batteries), are represented as boxes. The dashed lines outline the node points.

Look at Node A, which consists of points 1, 2, and 3. These points are really the same node or point connected by a zero-resistance wire. Similarly, four devices are connected at Node C.

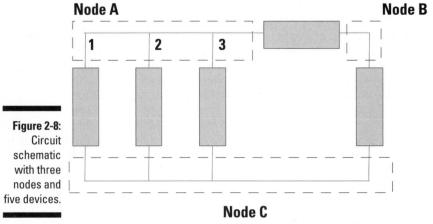

Figure 2-8:
Circuit schematic with three nodes and five devices.

Illustration by Wiley, Composition Services Graphics

Chapter 3

Exploring Simple Circuits with Kirchhoff's Laws

In This Chapter

▶ Discovering Kirchhoff's voltage law and current law

▶ Analyzing simple circuits with the help of Kirchhoff and Ohm

▶ Finding the equivalent resistance of series or parallel resistors and their combinations

*J*ust like you follow the law of gravity after jumping out of a perfectly good airplane to go skydiving, the devices (or elements) in any circuit have to follow certain laws. Whereas you follow the laws of nature, circuit elements such as resistors must follow Kirchhoff's laws.

This chapter introduces you to Gustav Kirchhoff's two circuit laws — Kirchhoff's voltage law (KVL) and Kirchhoff's current law (KCL) — and reveals how to use them in conjunction with Ohm's law, which I introduce in Chapter 2. With these three laws, you can solve for the current and voltage in any device in a circuit. The circuits in this chapter focus primarily on resistors driven by independent sources such as batteries.

Presenting Kirchhoff's Famous Circuit Laws

Gustav Kirchhoff's two laws — Kirchhoff's voltage law (KVL) and Kirchhoff's current law (KCL) — are essential for creating connection equations. *Connection equations* follow the energy and charge conservation laws when you connect devices (such as batteries or resistors) to form a circuit. Conservation laws tell you that electrons must behave in certain ways when you're connecting devices to form a circuit. In turn, the electron actions govern the behavior of the voltage and current around the loops and at the nodes.

More importantly, these connection equations don't depend on specific elements in the circuit. In other words, Kirchhoff's laws work no matter which connected devices you use in forming the circuit. The following sections get you acquainted with KVL and KCL and show you how to apply them to circuits. Rest assured that KVL and KCL are some of your best friends in the world of circuit analysis.

Kirchhoff's voltage law (KVL): Conservation of energy

Kirchhoff's voltage law (KVL) states that the algebraic sum of the voltages around a closed loop is zero at every instant. You can write this law as follows:

Voltage supplied (or delivered) = Voltage drops absorbed (or used up)

Electrical energy relates to *voltage,* the amount of energy required to move a given amount of charge (for more on voltage, see Chapter 2). So you can think of KVL as a mathematical representation of the law of conservation of energy. In a circuit, for example, a battery supplies power (a rate of energy) and a resistor dissipates the delivered power as heat (as when an incandescent light bulb emits heat). Another way of saying this is that the amount of supplied energy is equal to the amount of energy used up or absorbed.

To visualize KVL, suppose you're going for a walk. You start at your house and walk through your neighborhood, going up and down a number of hills. You head home using a different path, walking up and down another set of hills before ending your walk at home. Walking up and down the hills is analogous to voltage rises and drops. After starting and completing your walk, you form a closed path or loop ending at the starting point, and your net elevation (potential energy) doesn't change.

KVL and loops in a circuit go hand in hand. As you go around a loop, you enter and exit circuit devices. Two points are required to measure voltage, so if you enter and exit a device, you have enough points to find the device's voltage using KVL. After that, you can find the device's corresponding current by using relationships such as Ohm's law.

Formulating KVL expressions requires understanding the concept of voltage rises and drops. I get you acquainted with voltage rises and drops and show you how to calculate KVL equations in the next sections.

Identifying voltage rises and drops

Voltage rises occur when you go from a negative terminal to a positive terminal (– to +). They're usually associated with batteries or, more generally, *sources.* Not surprisingly, *voltage drops* occur when you go from a positive terminal to a negative terminal (+ to –), along the direction of electric current flow. They're commonly associated with passive devices, called *loads,* such as resistors.

Figure 3-1 shows you what voltage rises and drops look like in a circuit schematic. In Device 1, going along the direction of the current from the negative terminal to the positive terminal (left to right) results in a voltage rise. For Device 2, going from the positive terminal to the negative terminal (again, left to right) constitutes a voltage drop. As for the direction of the current, the *passive sign convention* tells you that the current flows from the positive terminal to the negative terminal. (For the scoop on the passive sign convention, turn to Chapter 2.)

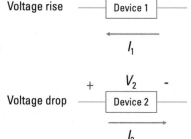

Voltage rise

V_1

Device 1

I_1

Voltage drop

V_2

Device 2

I_2

Figure 3-1:
A voltage
rise and
voltage
drop.

Illustration by Wiley, Composition Services Graphics

Always label your schematics appropriately so you know where the voltage terminals are and the direction of the current.

Forming a KVL equation

After labeling the voltage polarities (+ and –) in a circuit schematic, you can form the KVL equation. Simply choose your starting point and travel the path of the current through each device, noting the voltage as you enter each device. When you're back to where you started, add up all the rises and drops in voltage to get your KVL equation. Labeling the circuit appropriately helps you write your KVL equation correctly.

Consider the circuit in Figure 3-2. To account for the voltage rises and drops when going around the loop, you need to keep track of the voltage polarities and pick a *node* (a junction point where two or more devices are connected) to serve as both the starting and ending point. Then go around the loop in either a clockwise or counterclockwise direction. Also, make sure you're not passing through any device more than once for a single loop.

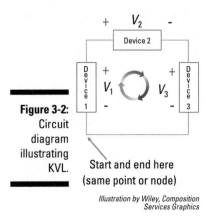

Figure 3-2:
Circuit
diagram
illustrating
KVL.

Start and end here
(same point or node)

Try to build the KVL equation for the circuit in Figure 3-2, starting at the lower-left corner of the diagram and going around the loop in a clockwise direction. The circuit consists of three loads. Keep account of voltage drops in the direction of the arrows. Because you enter Device 1 at its negative terminal and exit at its positive terminal, the resulting voltage rise as you go across Device 1 is $+V_1$. Then you enter Device 2 at its positive terminal and exit at its negative terminal. Consequently, the voltage drop across Device 2 can be expressed as $-V_2$. Next, you enter Device 3 at its positive terminal and exit at its negative terminal, so you wind up with a voltage of drop of $-V_3$. Finally, leaving Device 3 at the negative terminal takes you back to where you started. Now you have the information you need to create your KVL equation. When you add up the voltage rises and drops for all the devices, you get the following KVL equation:

$$V_1 - V_2 - V_3 = 0$$

You can get the same KVL equation for this example if you go counterclockwise and start at the upper-right corner of the circuit in Figure 3-2. In that case, the KVL equation would be

$$V_2 - V_1 + V_3 = 0$$

Algebraically, the preceding two equations are equivalent. Whether you go in a clockwise or counterclockwise direction, or whether you get voltage drops or rises for each device as you go around the loop, you obtain the same KVL equation.

If you want to show that the sum of the voltage rises is equal to the sum of the voltage drops (to reinforce that KVL is really a conservation-of-energy equation), write the KVL equation as follows:

Sum of voltage rises = Sum of voltage drops

Similarly, by using some algebra to isolate V_1 on one side of the equation, you get the following:

$$V_1 = V_2 + V_3$$

This form reinforces the fact that KVL is really like a conservation of energy equation.

Kirchhoff's current law (KCL): Conservation of charge

Kirchhoff's current law (KCL) tells you that the following is true at a node:

Sum of incoming currents = Sum of outgoing currents

So for all intents and purposes, KCL is really a mathematical representation of the conservation of charge. Think about it this way: When you apply voltage pressure using a battery, it supplies a current flow at one end, and the same amount of current is delivered to the rest of the circuit. Charge can't accumulate in the wire, which means the current flow is related to conservation of charge.

You can envision current — the flow of charges — as the flow of water. When you open a water faucet that has a hose connected to it, the amount of water flowing from the faucet is the same as the amount of water exiting at the other end of the hose. The water pressure from an external energy source creates the water flow, which means the water can't accumulate in the hose. Granted, this example illustrates conservation of mass rather than conservation of charge, but you get the idea.

KCL means that the current entering a node must equal the current going out of a node. Formally, KCL states that the algebraic sum of all the currents at a node is zero, but the in = out version is simpler because you don't have to figure out what's positive or negative for incoming or outgoing currents.

Tracking incoming and outgoing current

To create a KCL equation, you need to keep track of incoming currents and outgoing currents at each node. You're measuring a net current of zero for each node.

For practice following the ins and outs of currents, check out Figure 3-3, which shows three nodes. To minimize the clutter in the figure and show the consistency in the passive sign convention, I limit the voltage polarity notations (+ and –) to Devices 1 and 4. The current I_4 is flowing through Device 4.

I_4 is an outgoing current at Node A but an incoming current at Node B. This labeling convention holds true for the other devices and currents as well.

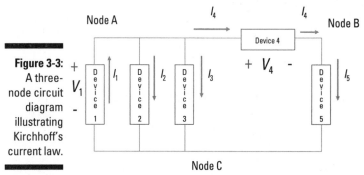

Figure 3-3:
A three-node circuit diagram illustrating Kirchhoff's current law.

Illustration by Wiley, Composition Services Graphics

Calculating KCL

When you have one incoming current and one outgoing current at a node, applying KCL is relatively straightforward. Consider Node B in Figure 3-3. The current entering Node B is I_4, and the current exiting Node B is I_5. Algebraically, the KCL equation for Node B is

$$\text{in} = \text{out} \quad \rightarrow \quad I_4 = I_5$$

Now consider Node A. The incoming current is I_1, and the outgoing currents are I_2, I_3, and I_4. The KCL equation for Node A is therefore

$$\text{in} = \text{out} \quad \rightarrow \quad I_1 = I_2 + I_3 + I_4$$

Finally, consider Node C. The incoming currents are I_2, I_3, and I_5, and the outgoing current is I_1. So the KCL equation for Node C is

$$\text{in} = \text{out} \quad \rightarrow \quad I_2 + I_3 + I_5 = I_1$$

The number of independent KCL equations you actually need is one fewer than the number of nodes for any circuit. If you want to save yourself a little time in the case of Figure 3-3, you can find the KCL equation for Node C simply by substituting the KCL equation for Node B into the KCL equation for Node A.

If you have a lot of devices connected at one node, label that node as your circuit reference point and count it as having 0 volts (I discuss the ground symbol and reference nodes in Chapter 2). Doing so will make your calculations cleaner when you start getting into more advanced circuit analysis. In Figure 3-3, Node C has the most devices connected to it, so that's the one you want to set equal to 0 volts as your circuit reference point.

Tackling Circuits with KVL, KCL, and Ohm's Law

Mathematically, you need two basic types of equations to analyze circuits:

- Device equations that describe the behavior between voltage and current for the component in question
- Connection equations derived from Kirchhoff's voltage and current laws (KVL and KCL) for any circuit

In the following sections, I show you how to apply these laws to the following types of circuits: a circuit with a battery and resistors, a series circuit, and a parallel circuit. In a series circuit, the same current flows through all the connected devices. In a parallel circuit, all the connected devices have the same voltage. Although the circuits differ, the general procedure is the same:

1. **Label the device terminals with the proper voltage polarities (+ and –) and voltage variables.**

2. **Assign the directions of the currents for the given circuit.**

   Apply the passive sign convention, with current flowing from the + sign to the – sign.

3. **Formulate KVL or KCL connection equations.**

4. **Apply device equations (such as Ohm's law for resistors) and then substitute the device equations into the connection equations.**

5. **Solve for the voltage and current for any device.**

Getting batteries and resistors to work together

The circuit in Figure 3-4 is made up of a battery (an ideal voltage source) and two resistors. Because the power supply is a source, the current direction is away from the positive terminal. As for the resistors, they follow the passive sign convention: The current flows from + to –.

To find the voltage and current for each device in Figure 3-4, you first need to take stock of the circuit. Because the circuit contains three devices, you have a total of six unknown voltages and currents.

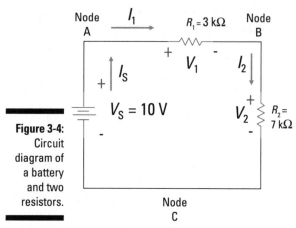

Figure 3-4:
Circuit
diagram of
a battery
and two
resistors.

Illustration by Wiley, Composition Services Graphics

Starting with voltage

You can begin your analysis with either of Kirchhoff's laws. In this example, go ahead and start with KVL.

You ultimately use both KVL and KCL because you need both of Kirchhoff's laws, along with device equations such as Ohm's law, to help you generate enough equations to solve for unknown voltages and currents. When you have a given number of voltages or currents to solve for — say, six in total — you need six independent equations. *Independent* means each equation can't be derived from the other equations. (If you can only get one equation by using the other equations, then that equation is *dependent*.) For each device, Ohm's law gives you one independent equation; the circuit in Figure 3-4 contains three devices, so you get three equations from Ohm's law. For each node, except for the reference or starting node, you get a KCL equation; this circuit has three nodes, so you get two KCL equations (not counting the ground Node C). You can get the dependent KCL equation for Node C from the other two KCL equations. Finally, this circuit gives you one independent KVL equation.

When formulating KVL, make sure you start and end at the same node. You wind up with the same KVL equation no matter where you start in the circuit, as long as you end at your starting point. For Figure 3-4, I start at the lower right-hand corner and move clockwise, producing the following KVL equation:

$$V_S - V_1 - V_2 = 0$$

Bringing in current

You have many unknowns with KVL and Ohm's law, so you need to establish more relationships between what's known and unknown. Use KCL next to get some more equations. Starting with Node A, the incoming current at Node A is I_s, and the outgoing current at Node A is I_1. So the KCL equation for Node A is

$$\text{in} = \text{out} \quad \rightarrow \quad I_s = I_1$$

For Node B, the incoming current is I_1, and the outgoing current is I_2. The KCL equation for Node B is

$$\text{in} = \text{out} \quad \rightarrow \quad I_1 = I_2$$

Combining the KCL equations for Nodes A and B, you have

$$I_s = I_1 = I_2$$

This is a neat equation because it says if you can figure out one of these currents, you've found the other two currents as well. For example, if you can find I_1, then you automatically know I_s and I_2.

Combining device equations with KVL

The voltage source V_s in Figure 3-4 is 10 volts. Plug this value into the previous KVL equation from the section "Starting with voltage":

$$V_s = 10 \text{ V} \rightarrow \qquad V_s = 10 \text{ V} = V_1 + V_2$$

Now you need device equations to figure out the unknown currents and voltages. To calculate the voltages for each device in Figure 3-4, use Ohm's law for resistors. You get two more equations relating voltage and current:

$$V_1 = I_1 R_1 = I_1 (3 \text{ k}\Omega) = I_1 (3{,}000 \ \Omega)$$

$$V_2 = I_2 R_2 = I_2 (7 \text{ k}\Omega) = I_1 (7{,}000 \ \Omega)$$

Note that you can write both equations in terms of I_1 because $I_1 = I_2$.

Now substitute the Ohm's law values of V_1 and V_2 into $V_s = 10 \text{ V} = V_1 + V_2$:

$$10 \text{ V} = (3{,}000 \ \Omega) I_1 + (7{,}000 \ \Omega) I_1 = (10{,}000 \ \Omega) I_1$$

Solving for I_1 gives you

$$I_1 = \frac{10 \text{ V}}{10{,}000 \ \Omega} = 0.001 \text{ A} = 1 \text{ mA}$$

Because $I_2 = I_s = I_1 = 0.001$ A, you know the current for all three devices.

Finally, you can plug in the values to figure out the voltages for the three devices using Ohm's law:

$$V_1 = I_1(3,000 \ \Omega) = (0.001 \ \text{A})(3,000 \ \Omega) = 3 \ \text{V}$$

$$V_2 = I_1(7,000 \ \Omega) = (0.001 \ \text{A})(7,000 \ \Omega) = 7 \ \text{V}$$

Verify the KVL equation by substituting in the voltages to show that the total sum of the voltages is equal to zero. It's good practice to check your results.

Note that the 10-volt power supply is divided proportionally between the two resistors of 3 volts and 7 volts.

Summarizing the results

To give you some insight into the calculations in this section, the following table lists the voltage and current for each device in the circuit. It also shows the power supplied by the voltage source and the power dissipated by the resistor using $P = IV$, where P is the power supplied by a source or absorbed by a load device, V is the voltage across the device, and I is the current through the device. *Note:* You don't need to develop a similar table for each circuit you analyze; this table just serves to illustrate some points.

Device	Current (I)	Voltage (V)	Power (W)
V_s	–1 mA	10 V	–100 mW
R_1	1 mA	3 V	30 mW
R_2	1 mA	7 V	70 mW

As you can see from the table, you have the same amount of current flowing through each device, which tells you that the circuit in question is a series circuit. As for the power, remember that negative power is supplied power and that positive power is dissipated or absorbed power; therefore, the power supplied by the battery V_s is equal to the sum of the power dissipated or absorbed by the two resistors, which illustrates the conservation of energy.

Sharing the same current in series circuits

Two devices or elements are connected *in series* when they have one common node where no other devices have currents flowing through them. In other words, the same current flows through each device, and the current can only flow forward. It basically has a one-way ticket to ride with no alternative routes.

If you're having trouble picturing series circuits, imagine that you've just bought a brand-new blanket that does a good job of trapping body heat. You also have a thin blanket, but it's a poor heat insulator. You know it's going to be a cold night, so you're trying to figure out how best to keep warm. Should the new blanket go on top or bottom of the old blanket? If your gut feeling is that it shouldn't matter, you're correct. Why? Because the blankets are connected in series. Heat must go through both blankets before it escapes. Just like heat flows from hot to cold places, current flows from high (+) to low (−) voltage. The blankets are heat insulators in series behaving like resistors connected in series. If the order of the resistors changes, you still get the same amount of current going through each of them.

Figure 3-5 shows three resistors connected in series and three resistors that aren't connected in series. You can see that the same current *I* flows through each of the three resistors in the series circuit.

Resistors connected in series

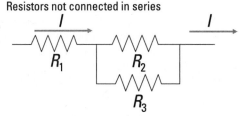

$R_1 = 1k\Omega$ $R_2 = 2k\Omega$ $R_3 = 3k\Omega$

I

Resistors not connected in series

I   *I*

R_1 R_2

R_3

Figure 3-5: Resistors connected in series (top) and not connected in series (bottom).

Illustration by Wiley, Composition Services Graphics

In the bottom circuit, the current *I* through R_1 splits proportionally between R_2 and R_3. Intuitively, if R_2 and R_3 have the same resistance value, then the current *I* splits in half: $I/2$. If R_2 has a bigger resistance value than R_3, then the current through R_2 will have a smaller value than the current through R_3. Either way, the currents flowing out between R_2 and R_3 will add up to the same current *I*.

The voltage across the three resistors connected in series is given as *V*, and the voltage for each of the resistors R_1, R_2, and R_3 is V_1, V_2, and V_3, respectively. Now apply Ohm's law, $V = IR$, for the voltages. Because the current is the same for each resistor in a series connection, this implies that

$$V = V_1 + V_2 + V_3$$
$$= IR_1 + IR_2 + IR_3$$
$$= I\left(R_1 + R_2 + R_3\right)$$
$$= IR_T$$

where R_T is the total resistance given as $R_T = R_1 + R_2 + R_3$. The preceding equation is a form of Ohm's law. So in Figure 3-5, the total resistance is $R_T = 1 \text{ k}\Omega + 2 \text{ k}\Omega + 3 \text{ k}\Omega = 6 \text{ k}\Omega$.

For resistors connected in series, the total resistance is simply the sum of the resistances. Whenever you see two or more resistors connected to a series circuit, you can replace the individual resistances with the total equivalent resistance for the series circuit — a tactic that comes in handy when you want to simplify a circuit for analysis. Similarly, you can break up a single resistor into smaller resistors that add up to the value of the single resistor. And don't forget that the same current flows through each series resistor.

When you check your answer, remember that the total resistance for resistors connected in series is always greater than the value of any one resistor.

Climbing the ladder with parallel circuits

Connected devices have the same voltage when they're connected in parallel. The lights on a string of Christmas lights are connected in parallel, as are all major appliances in a house. Figure 3-6 illustrates a parallel circuit that consists of three devices.

You can tell when you're looking at a parallel circuit because the circuit diagram looks like a ladder lying on its side. You can also say that devices are connected in parallel when they form a loop that doesn't encircle any other elements. Devices are connected in parallel when they have two nodes in common.

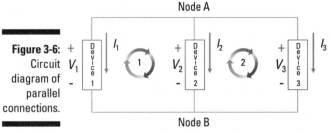

Node A

Figure 3-6: Circuit diagram of parallel connections.

Node B

Illustration by Wiley, Composition Services Graphics

To start analyzing this circuit, first formulate a KVL equation for Loop 1. You can start anywhere in the circuit, but I've started at the lower-left corner, which gives me the following equation:

$$V_1 - V_2 = 0 \quad \rightarrow \quad V_1 = V_2$$

For Loop 2, start from the lower-left corner of Loop 2, giving you another a KVL equation:

$$V_2 - V_3 = 0 \quad \rightarrow \quad V_3 = V_2 \quad \rightarrow \quad V_1 = V_2 = V_3$$

You now know that with the circuit configuration in Figure 3-6, the KVL analysis yields the same voltage for each device given in the circuit.

Next, apply KCL for Node A, where there are no incoming currents and where outgoing currents consist of I_1, I_2, and I_3. The result?

$$\text{in = out} \quad \rightarrow \quad 0 = I_1 + I_2 + I_3$$

If you suppose that the devices are resistors — R_1, R_2, and R_3 — then you can use Ohm's law to find the following device equations:

$$I_1 = \frac{V}{R_1} \quad , \qquad I_2 = \frac{V}{R_2} \quad , \qquad I_3 = \frac{V}{R_3}$$

When you substitute Ohm's law into the KCL equation, $0 = I_1 + I_2 + I_3$, you get the following (where R_T is the total equivalent resistance for these three resistors connected in parallel):

$$0 = \frac{V}{R_1} + \frac{V}{R_2} + \frac{V}{R_3}$$
$$0 = \frac{1}{R_1} + \frac{1}{R_2} + \frac{1}{R_3} = \frac{1}{R_T}$$

I simplified that last equation by dividing both sides of the equation by V. As you can see, finding the total resistance for resistors connected in parallel is a bit more complicated than finding the total resistance for a series circuit, which I explain how to do in the preceding section.

Describing total resistance using conductance

For more than two resistors connected in parallel, you can get an alternate description of the total resistance by using the definition of *conductance*, $G = 1/R$, which I introduce in Chapter 2. Using $G = 1/R$ for Figure 3-6 gives you the following equation to find the total conductance (which is simply the sum of the conductances for each device that's connected in parallel):

$$\frac{1}{R_T} = \frac{1}{R_1} + \frac{1}{R_2} + \frac{1}{R_3} \quad \rightarrow \quad G_T = G_1 + G_2 + G_3$$

Try your hand at using the definition of conductance to find the total resistance for the circuit in Figure 3-7.

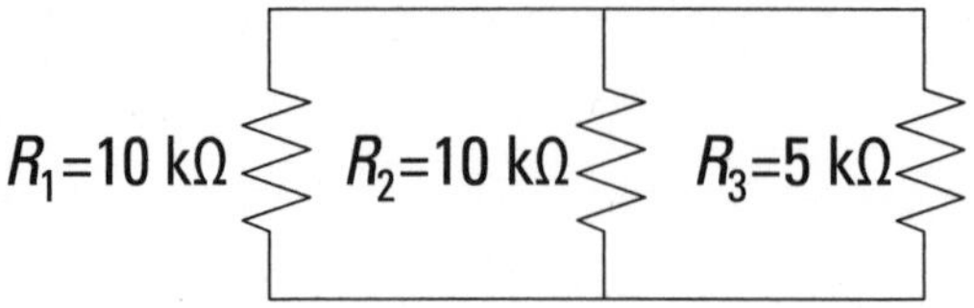

Figure 3-7:
Parallel
connections
of resistors.

$R_1 = 10 \text{ k}\Omega$ $R_2 = 10 \text{ k}\Omega$ $R_3 = 5 \text{ k}\Omega$

Illustration by Wiley, Composition Services Graphics

Here's the calculation of the total resistance given the three resistors:

$$\frac{1}{R_T} = \frac{1}{10 \text{ k}\Omega} + \frac{1}{10 \text{ k}\Omega} + \frac{1}{5 \text{ k}\Omega}$$

$$G_T = 0.1 \text{ mS} + 0.1 \text{ mS} + 0.2 \text{ mS} = 0.4 \text{ mS}$$

$$R_T = \frac{1}{0.4 \text{ mS}} = 2.5 \text{ k}\Omega$$

Using a shortcut for two resistors in parallel

Two resistors connected in parallel commonly appear in many circuits, so it's convenient to have a simple formula for this case. In words, the total resistance for two parallel resistors is the product of the two resistors divided by the sum of the two resistors. Algebraically, that looks like the following:

$$\frac{1}{R_T} = \frac{1}{R_1} + \frac{1}{R_2} \quad \rightarrow \quad R_T = \frac{R_1 R_2}{R_1 + R_2}$$

Finding equivalent resistor combinations

To better understand equivalent resistances, try calculating the total equivalent resistance of the three parallel resistors in Figure 3-7. You can find the equivalent resistance in pairs, as in Circuit A of Figure 3-8. Using the preceding equation for pairs of resistors yields

$$\frac{1}{R_{eq1}} = \frac{1}{10 \text{ k}\Omega} + \frac{1}{10 \text{ k}\Omega} \quad \rightarrow \quad R_{eq1} = \frac{10 \text{ k}\Omega \cdot 10 \text{ k}\Omega}{10 \text{ k}\Omega + 10 \text{ k}\Omega} = 5 \text{ k}\Omega$$

The two 10-kΩ resistors in parallel in Circuit A are replaced with a 5-kΩ resistor in Circuit B, where you now have two 5-kΩ resistors connected in parallel. Transform the circuit again, replacing these two resistors in parallel with their equivalent:

$$\frac{1}{R_{eq2}} = \frac{1}{5 \text{ k}\Omega} + \frac{1}{5 \text{ k}\Omega} \quad \rightarrow \quad R_{eq2} = \frac{5 \text{ k}\Omega \cdot 5 \text{ k}\Omega}{5 \text{ k}\Omega + 5 \text{ k}\Omega} = 2.5 \text{ k}\Omega$$

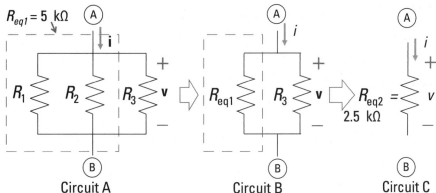

Figure 3-8:
Circuit combining series and parallel resistors.

$R_1 = R_2 = 10 \text{ k}\Omega$

$R_3 = 5 \text{ k}\Omega$

$R_{eq1} = \dfrac{R_1 R_2}{R_1 + R_2} = \dfrac{10 \text{ k}\Omega \cdot 10 \text{ k}\Omega}{10 \text{ k}\Omega + 10 \text{ k}\Omega} = 5 \text{ k}\Omega \longrightarrow R_{eq2} = \dfrac{R_{eq1} \cdot R_3}{R_{eq1} + R_3} = \dfrac{5 \text{ k}\Omega \cdot 5 \text{ k}\Omega}{5 \text{ k}\Omega + 5 \text{ k}\Omega} = 2.5 \text{ k}\Omega$

Illustration by Wiley, Composition Services Graphics

Circuit B is transformed to Circuit C, which has an equivalent resistance of 2.5 kΩ.

This example shows that if you want to cut the value of a particular resistor in half, you can connect two equal resistors in parallel. Circuits A, B, and C are all equivalent circuits because they have the same voltage v across Terminals A and B and the same *net* current i going through the resistor network.

Finding the faulty bulb in a string of Christmas lights

Early Christmas lights were frustrating for reasons beyond the tangled mess they created if you weren't careful. The individual light bulbs were connected in series, so when one bulb went out, none of the bulbs would light. It was difficult to determine which bulb had burnt out because there wasn't a current going through each of the light bulbs. You had to check every single bulb to find the culprit.

Fortunately, manufacturers saw the (multicolored, twinkling) light. Most of today's Christmas lights are connected in parallel. So when one light bulb is bad, the other lights stay on because they all share the same voltage — making it easy to find the one faulty bulb you need to replace. As for the tangled mess, even electrical engineers haven't solved that one!

Combining series and parallel resistors

You can use the concepts of series and parallel resistors to transform a complex circuit into a simpler circuit. Replacing part of a complicated-looking circuit with a simpler but equivalent circuit simplifies the math.

Figure 3-9 shows a complex circuit with resistors connected in series and in parallel. Your job is to find the total resistance so you can replace all those resistors with a single resistor.

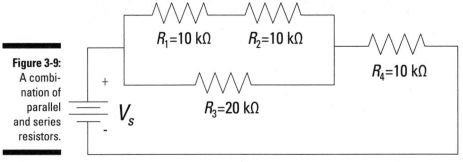

Figure 3-9: A combination of parallel and series resistors.

Illustration by Wiley, Composition Services Graphics

In Figure 3-9, resistors R_1 and R_2 are connected in series. This series combination is equivalent to

$$R_{eq1} = R_1 + R_2 = 10 \text{ k}\Omega + 10 \text{ k}\Omega = 20 \text{ k}\Omega$$

R_{eq1} is connected in parallel with R_3. You calculate the equivalent resistance value for this parallel combination as

$$R_{eq2} = \frac{R_{eq1}R_3}{R_{eq1} + R_3} = \frac{(20 \text{ k}\Omega)(20 \text{ k}\Omega)}{20 \text{ k}\Omega + 20 \text{ k}\Omega} = 10 \text{ k}\Omega$$

R_{eq2} is in series with R_4. This series combination yields

$$R_{eq} = R_{eq2} + R_4 = 10 \text{ k}\Omega + 10 \text{ k}\Omega = 20 \text{ k}\Omega$$

Chapter 4

Simplifying Circuit Analysis with Source Transformation and Division Techniques

---

In This Chapter

▶ Recognizing equivalent circuits

▶ Transforming circuits into equivalent series and parallel circuits

▶ Analyzing circuits with voltage and current divider techniques

---

*U*sing Kirchhoff's laws and Ohm's laws (see Chapter 3) can get pretty laborious when you're analyzing complex circuits. Fortunately, you can make analyzing circuits easier by replacing part of the circuit with a simpler but equivalent circuit.

Through a *makeover* or *transformation technique,* you modify a complex circuit so that in the transformed circuit, the devices are all connected in series or in parallel. After the transformation, you no longer need to systematically apply Kirchhoff's laws, because you can use shortcuts: the current divider technique and the voltage divider technique.

In this chapter, I explain how to make the transformation and apply both types of divider techniques. Rest assured that the info in this chapter can make your life a little easier when you start analyzing more-complex circuits.

Equivalent Circuits: Preparing for the Transformation

When you're analyzing a complex circuit, you can simplify the math by replacing part of the circuit with a simpler, equivalent circuit. Two circuits are said to be *equivalent* if they have the same *i-v* characteristics at a pair of terminal connections. (You can find information about the *i-v* characteristics of various electrical devices in Chapter 2.)

You find the *i-v* characteristic for each circuit by using Kirchhoff's laws and Ohm's law, which give you the equations that relate the current *i* and voltage *v* across two terminals (see Chapter 3 for details). Then you compare the *i-v* relationships associated with the pair of terminals to find out in which conditions the circuits are equivalent. Even better, after you understand how to do source transformations, you no longer need to rely completely on Kirchhoff's and Ohm's laws to complete your analysis.

Take a look at the practical models of independent voltage and current sources in Figure 4-1. Circuit A depicts an ideal voltage source connected in series with a resistor, and Circuit B depicts an ideal current source connected in parallel with a resistor. In the following example, I show you that these two circuits are considered equivalent because they have the same *i-v* characteristics at the terminal pair A and B.

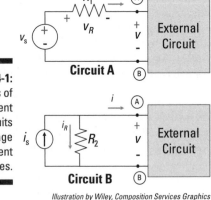

Figure 4-1:
Models of
equivalent
circuits
with voltage
and current
sources.

Illustration by Wiley, Composition Services Graphics

To find the *i-v* characteristic of Circuit A, you have to develop the relationship between the current *i* and voltage *v* for Terminals A and B. You do this by using Kirchhoff's and Ohm's laws.

Kirchhoff's voltage law (KVL) says that the sum of the voltage drops and rises around a loop is zero. In other words, the voltage source has to equal the voltage drops across the resistors. Therefore, using KVL for Circuit A produces

$$v_s = v_R + v$$

Using Ohm's law for resistor R_1 gives you the following voltage:

$$v_R = iR_1$$

Substituting the value of v_R into $v_s = v_R + v$ yields

$$v_s = R_1 i + v$$

One way to get the *i-v* characteristic for Circuit A is to solve for v, which yields the following:

$$v = v_s - R_1 i$$

The resulting equation relates the voltage v and the current i at Terminals A and B in Circuit A of Figure 4-1. If you know the current and voltage from the input voltage source, you can find the voltage output.

An alternate form of the *i-v* characteristic requires manipulating $v = v_s - R_1 i$ by solving for current i to obtain $i = \dfrac{v_s}{R_1} - \dfrac{v}{R_1}$. When you do this, knowing the voltage input provides you with the current output at Terminals A and B.

Look at Circuit B in Figure 4-1 to find a similar *i-v* relationship at Terminals A and B. You use Kirchhoff's current law (KCL), which states that the sum of the incoming currents is equal to the sum of the outgoing currents at any node or terminal — here, at Terminal A or Terminal B. KCL yields

$$i_s = i_R + i$$

Using Ohm's law for R_2 gives you the following:

$$i_R = \frac{v}{R_2}$$

When you substitute the value of i_R into $i_s = i_R + i$, you get

$$i_s = \frac{v}{R_2} + i$$

Solving for v gives you the following *i-v* characteristic:

$$v = R_2 i_s - R_2 i$$

This equation relates the voltage v and the current i at Terminals A and B for Circuit B.

Now you can compare the result for Circuit B, $v = R_2 i_s - R_2 i$, with the result for Circuit A, $v = v_s - R_1 i$, to find the conditions for equivalent circuits. One way of doing so is to equate v between the two circuits. For this example, this approach gives you:

$$R_2 i_s - R_2 i = v_s - R_1 i$$

If you rearrange these equations to group the independent sources and collect like terms for the current i, you wind up with the following conditions:

$$\underbrace{(R_2 i_s - v_s)}_{=0} + \underbrace{(R_1 i - R_2 i)}_{=0} = 0$$

The first expression in parentheses deals with the independent sources, and the second collects like terms with current i. For this equation to be equal to zero, you set the terms in parentheses equal to zero, which gives you the following two equations:

$$R_1 = R_2 = R$$
$$v_s = R_2 i_s = i_s R$$

Because R_1 and R_2 are equal to each other, removing their subscripts yields a general resistor value of R for the two circuits. These are the conditions in which the two circuits are said to be equivalent.

Transforming Sources in Circuits

Each device in a series circuit has the same current, and each device in a parallel circuit has the same voltage. Therefore, finding the current in each device in a circuit is easier when the devices are all connected in parallel, and finding the voltage is easier when they're all connected in series. Through a circuit *transformation,* or *makeover,* you can treat a complex circuit as though all its devices were arranged the same way — in parallel or in series — by appropriately changing the independent source to either a current or voltage source.

Changing the practical voltage source to an equivalent current source (or vice versa) requires the following conditions (see the preceding section to find out why these conditions characterize equivalent circuits):

 ✔ The resistors must be equal in both circuits.

 ✔ The source transformation must be constrained by $v_s = i_s R$.

The constraining equation, $v_s = i_s R$, looks like Ohm's law, which should help you remember what to do when transforming between independent voltage and current sources.

In this section, I show you how to transform a circuit.

Converting to a parallel circuit with a current source

Transformation techniques let you convert a practical voltage source with a resistor connected in series to a current source with a resistor connected in parallel. Therefore, you can convert a relatively complex circuit to an equivalent circuit if all the devices in the external circuit are connected in parallel. You can then find the current of individual devices by applying the current divider techniques that I discuss later in "Cutting to the Chase Using the Current Divider Technique."

When switching from a voltage source to a current source, the resistors have to be equal in both circuits, and the source transformation must be constrained by $v_s = i_s R$. Solving the constraint equation for i_s allows you to algebraically convert the practical voltage source into a current source:

$$i_s = \frac{v_s}{R}$$

Figure 4-2 illustrates the conversion of a voltage source, in Circuit A, into an equivalent current source, in Circuit B. The resistors, R, are equal, and the constraint equation was applied to change the voltage source into a current source.

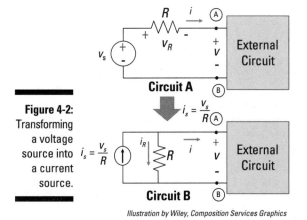

Figure 4-2:
Transforming a voltage source into a current source.

Illustration by Wiley, Composition Services Graphics

Figure 4-3 shows the conversion with some numbers plugged in. Both circuits contain the same 3-kΩ resistor, and the source voltage in Circuit A is 15 volts. With this information, you can find the source current, i_s, for the transformed Circuit B.

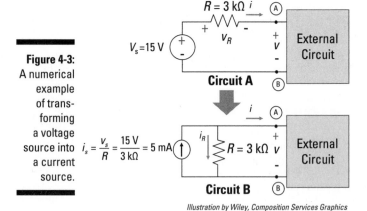

Figure 4-3:
A numerical example of transforming a voltage source into a current source.

Illustration by Wiley, Composition Services Graphics

Use the constraint equation to find the source current in Circuit B. Here's what you get when you plug in the numbers:

$$i_s = \frac{v_s}{R} = \frac{15 \text{ V}}{3 \text{ k}\Omega} = 5 \text{ mA}$$

Changing to a series circuit with a voltage source

You can convert a current source connected in parallel with a resistor to a voltage source connected in series with a resistor. You use this technique to form an equivalent circuit when the external circuit has devices connected in series.

Converting a practical current source connected with a resistor in parallel to a voltage source connected with a resistor in series follows the conditions for equivalent circuits:

✔ The resistors must be equal in both circuits.

✔ The source transformation must be constrained by $v_s = i_s R$.

Figure 4-4 illustrates how to convert a current source into a voltage source.

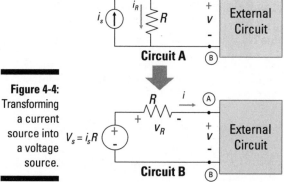

Figure 4-4:
Transforming a current source into a voltage source.

Illustration by Wiley, Composition Services Graphics

Figure 4-5 depicts the same transformation of a current source to a voltage source with some numbers plugged in. Both circuits contain the same 3-kΩ resistor, and the current source in Circuit A is 5 mA.

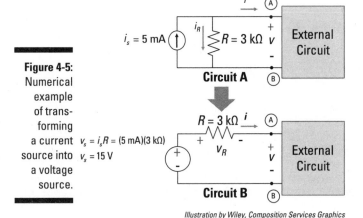

Figure 4-5:
Numerical
example
of trans-
forming
a current
source into
a voltage
source.

You can use the constraint equation to find the source voltage for Circuit B. Plugging in the numbers produces the following:

$$v_s = i_s R = (5 \text{ mA})(3 \text{ k}\Omega) = 15 \text{ V}$$

Suppose you have a complex circuit that has a current source, a resistor connected in parallel, and an external circuit with multiple resistors connected in series. You can transform the circuit so that it has a voltage source connected with all the resistors in series.

Consider Circuit A in Figure 4-6, where the right side of Terminals A and B consists of two resistors connected in series. On the left side of Terminals A and B is a practical current source modeled as an ideal current source in parallel with a resistor.

You want all the devices to be connected in series, so you need to move R when you transform the circuit. To transform the circuit, change the current source to a voltage source and move R so that it's connected in series rather than in parallel. When you use the constraint equation $v_s = i_s R$ to find the source voltage, remember that R is the resistor you moved.

Circuit B is a series circuit where all the devices share the same current. You can find the voltage through R, R_1, and R_2 using voltage divider techniques, which I discuss in the next section.

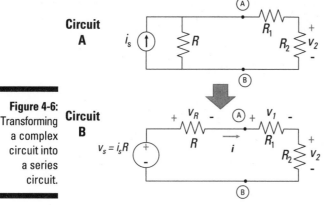

Figure 4-6:
Transforming
a complex
circuit into
a series
circuit.

Divvying It Up with the Voltage Divider

The *voltage divider* technique allows you to calculate the voltage for each device connected in series with an input voltage source. In the preceding section, I show you how to transform a circuit to a series circuit with a voltage source. This section shows you how to formulate the voltage divider equation. Then you see the voltage divider equation at work.

If a circuit problem with a current source asks you to find the voltage for a particular device, you may find it easier to convert the circuit to a series circuit first. Then you can find voltage using the voltage divider technique.

Getting a voltage divider equation for a series circuit

You use the voltage divider when the device in the circuit is connected in series and is driven by a voltage source. The input voltage source is divided proportionally according to the resistor values.

To get the voltage divider equation, you start with the fact that for a series circuit, the same current flows through each resistor. With this current, you use Kirchhoff's voltage law (KVL) and Ohm's law to obtain the voltage across a particular resistor. You can solve for (or eliminate) the current in the expression. In the resulting equation, the desired voltage across the resistor is proportional to the input source voltage. Because the voltage source is multiplied by a ratio of resistors having a value of less than 1, the desired output and device voltage are always less than the input source voltage.

Look at Circuit B of Figure 4-6, which has a voltage source and three resistors connected in series. You want to calculate the voltage across the resistors.

KVL says that the sum of the voltage rises and drops around a loop is equal to zero. So applying KVL to Circuit B produces the following:

$$v_s = v_R + v_1 + v_2$$

Because the circuit is connected in series, the same current i flows through each of the resistors. Using Ohm's law for each resistor yields

$$v_R = iR$$
$$v_1 = iR_1$$
$$v_2 = iR_2$$

Substituting v_R, v_1, and v_2 into $v_s = v_R + v_1 + v_2$ and factoring out the current i gives you

$$v_s = i(R + R_1 + R_2)$$

When you divide the voltage across Resistor 1, which is $v_1 = iR_1$, by $v_s = i(R + R_1 + R_2)$, you get one form of the desired voltage divider equation:

$$\frac{v_1}{v_s} = \frac{iR_1}{i(R + R_1 + R_2)}$$
$$\frac{v_1}{v_s} = \frac{R_1}{R + R_1 + R_2}$$

This form of the voltage divider is often referred as a *voltage transfer function*, which relates the output voltage (voltage v_1 for this example) to the input voltage source (which is v_s in this case). You can find the output voltage if you know the input voltage source.

Solving for v_1 yields the following voltage divider equation:

$$v_1 = v_s\left(\frac{R_1}{R + R_1 + R_2}\right)$$

You can find similar voltage divider relationships for v_2 and v_R:

$$v_2 = v_s \left(\frac{R_2}{R + R_1 + R_2} \right)$$

$$v_R = v_s \left(\frac{R}{R + R_1 + R_2} \right)$$

These divider equations show that to find the voltage across a particular resistor, you simply multiply the input source voltage by the desired resistor and divide by the total resistance of the series circuit. That is, the voltage of each device of Figure 4-6 depends on the ratio of resistors multiplied by the source voltage.

The voltage across each resistor is always less than and proportional to the supplied independent source voltage because the ratio of resistors is always less than 1. This idea offers a handy check of your calculations. The largest voltage goes across the largest resistor, and the lowest voltage goes across the smallest resistor.

Figuring out voltages for a series circuit with two or more resistors

Voltage divider techniques work well for a series circuit that has two or more resistors. You calculate the output voltage by multiplying the input source voltage by the desired resistor and dividing by the total resistance in the circuit.

I use Figure 4-7 to illustrate the voltage divider technique numerically. The given Circuit A has a source current of 5 milliamps as well as a 4-kΩ resistor arranged in parallel with a series combination of 6-kΩ and 10-kΩ resistors. To find the voltage across the resistors, you first transform Circuit A so that it has a voltage source and all three resistors in series.

Start by finding the source voltage in the transformed circuit, Circuit B. The transformation must be constrained by $v_s = i_s R$, so here's the source voltage:

$$v_s = i_s R = (5 \text{ mA})(4 \text{ k}\Omega) = 20 \text{ V}$$

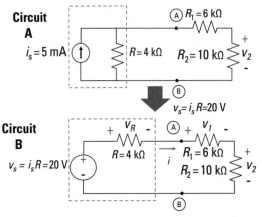

Illustration by Wiley, Composition Services Graphics

Figure 4-7: Numerical example of the voltage divider method.

According to the voltage divider equation, you find the voltage across a resistor by multiplying the source voltage by the desired resistor and then dividing by the total resistance of the series circuit (see the preceding section for details). Try calculating the voltage for each resistor shown. Use the voltage divider shortcut and plug in the numbers:

$$v_1 = 20 \text{ V} \cdot \left(\frac{6 \text{ k}\Omega}{4 \text{ k}\Omega + 6 \text{ k}\Omega + 10 \text{ k}\Omega} \right) = 6 \text{ V}$$

$$v_2 = 20 \text{ V} \cdot \left(\frac{10 \text{ k}\Omega}{4 \text{ k}\Omega + 6 \text{ k}\Omega + 10 \text{k}\Omega} \right) = 10 \text{ V}$$

$$v_R = 20 \text{ V} \cdot \left(\frac{4 \text{ k}\Omega}{4 \text{ k}\Omega + 6 \text{ k}\Omega + 10 \text{ k}\Omega} \right) = 4 \text{ V}$$

Finding voltages when you have multiple current sources

Analyzing a circuit that has multiple current sources and parallel resistors would be tedious if you could only use Kirchhoff's laws and Ohm's law. However, thanks to the power of source transformation and the voltage divider technique, the analysis is relatively straightforward.

Circuit A in Figure 4-8 has two current sources and two parallel resistors. What's the voltage, v_1, through resistor R_1?

You transform this circuit in two stages. First transform the circuit so that it has two voltage sources and all the resistors arranged in series. Then combine the voltage sources to get one equivalent voltage source. After that, you can find v_1, the voltage across R_1, using the voltage divider technique.

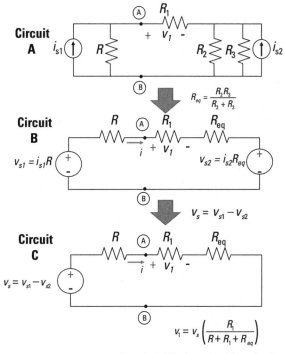

Figure 4-8:
Circuit
analysis $V_s = V_{s1} - V_{s2}$
with mul-
tiple current
sources.

Illustration by Wiley, Composition Services Graphics

Circuit B of Figure 4-8 is the transformation of Circuit A using several operations. On the right side of Circuit A, you want to move both R_2 and R_3 so that they're connected in series. These resistors are connected in parallel, so find their equivalent resistance, R_{eq}:

$$R_{eq} = \frac{R_2 R_3}{R_2 + R_3}$$

Next, transform the two current sources by converting them to voltage sources:

$$v_{s1} = i_{s1}R$$
$$v_{s2} = i_{s2}R_{eq}$$

Circuit C of Figure 4-8 completes the transformation. The two voltage sources connected in series are combined, forming one equivalent voltage source. Observing the voltage polarities results in one voltage source, as follows:

$$v_s = v_{s1} - v_{s2}$$

For a numerical example, check out the circuit in Figure 4-9. Your goal is to find v_1, the voltage of the 11-kΩ resistor. Start the circuit transformation by putting the resistors in series and switching to voltage sources.

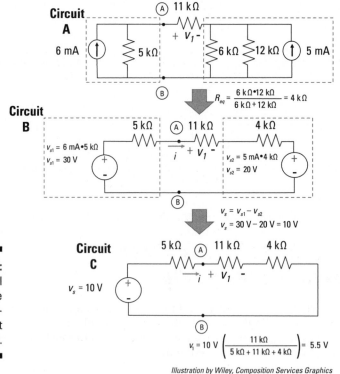

Illustration by Wiley, Composition Services Graphics

Figure 4-9: Numerical example with multiple current sources.

Here's the equivalent total resistance for two parallel resistors of 6 kΩ and 12 kΩ:

$$R_{eq} = \frac{6 \text{ k}\Omega \cdot 12 \text{ k}\Omega}{6 \text{ k}\Omega + 12 \text{ k}\Omega} = 4 \text{ k}\Omega$$

Use the transformation equations $v_{s1} = i_{s1}R$ and $v_{s2} = i_{s2}R_{eq}$ to convert the two current sources to two voltage sources:

$$v_{s1} = 6 \text{ mA} \cdot 5 \text{ k}\Omega = 30 \text{ V}$$
$$v_{s2} = 5 \text{ mA} \cdot 4 \text{ k}\Omega = 20 \text{ V}$$

You can see the source transformations and equivalent resistance in Circuit B of Figure 4-9.

The voltage sources are connected in series, so combine them, noting their polarities:

$$v_s = v_{s1} - v_{s2} = 30 \text{ V} - 20 \text{ V} = 10 \text{ V}$$

Circuit C of Figure 4-9 shows the completed and simplified transformation. Now you can use the voltage divider equation to find v_1:

$$v_1 = 10 \text{ V} \cdot \frac{11 \text{ k}\Omega}{4 \text{ k}\Omega + 11 \text{ k}\Omega + 5 \text{ k}\Omega} = 5.5 \text{ V}$$

Even though the voltage divider shortcut for series circuits lets you find an unknown voltage without using Kirchhoff's and Ohm's laws, this technique was developed from the foundational equations of Kirchhoff's and Ohm's laws.

Using the voltage divider technique repeatedly

When a part of a circuit has a combination of series and parallel resistors, you can use the voltage divider technique repeatedly. For example, you may use this approach when resistors are connected in parallel and one of the parallel branches has a series combination.

Consider Circuit A in Figure 4-10. You could find all the components' voltages and currents using Kirchhoff's voltage and current laws. But if all you want is the voltage across a specific device, you can take a shortcut with source transformation and the voltage divider technique. You can find the voltage v_x across the 3-kΩ resistor using the voltage divider technique repeatedly.

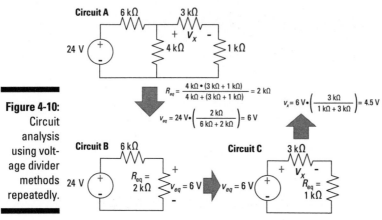

Figure 4-10:
Circuit analysis using voltage divider methods repeatedly.

On the right side of Circuit A, you have a resistor series combination of 3 kΩ and 1 kΩ connected in parallel with the 4-kΩ resistor. The total resistance for this combination is

$$R_{eq} = \frac{4\ \text{k}\Omega \cdot (3\ \text{k}\Omega + 1\ \text{k}\Omega)}{4\ \text{k}\Omega + (3\ \text{k}\Omega + 1\ \text{k}\Omega)} = 2\ \text{k}\Omega$$

I tell you how to find equivalent resistance in Chapter 3.

Circuit B shows the circuit after you combine these resistors. Calculate the voltage across R_{eq} using the voltage divider method:

$$v_{eq} = 24\ \text{V} \cdot \frac{2\ \text{k}\Omega}{6\ \text{k}\Omega + 2\ \text{k}\Omega} = 6\ \text{V}$$

However, the 6 volts also go across the resistor series combination of 3 kΩ and 1 kΩ, as depicted in Circuit C of Figure 4-10. Use the voltage divider method once again to find v_x:

$$v_x = 6\ \text{V} \cdot \frac{3\ \text{k}\Omega}{3\ \text{k}\Omega + 1\ \text{k}\Omega} = 4.5\ \text{V}$$

Cutting to the Chase Using the Current Divider Technique

The *current divider* technique lets you easily calculate the current for each device connected in parallel when the devices are driven by an input current source. Earlier in "Converting to a parallel circuit with a current source," I show you how to transform your circuit. This section shows you where the current divider equation comes from and how to apply it.

Getting a current divider equation for a parallel circuit

For devices connected in parallel with a current source, the current divider technique allows you to find the current through each device. Basically, you're looking at how the current source distributes its supplied current to each device, depending on the ratio of conductances (or resistances) in the circuit.

The current divider shortcut replaces using Kirchhoff's current law and Ohm's law in finding the current through each device. Of course, the shortcut works only because it's based on these fundamental laws. To see where the current divider equation comes from, look at Figure 4-11. Circuit A is a complex circuit with a voltage source. You want to find an equation to calculate the current through each resistor.

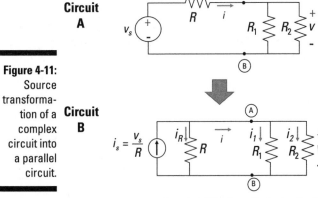

Figure 4-11:
Source transformation of a complex circuit into a parallel circuit.

Illustration by Wiley, Composition Services Graphics

Start by transforming the circuit. Circuit B is Circuit A transformed into a parallel circuit with a current source. Kirchhoff's current law (KCL) says that the sum of the incoming currents is equal to the sum of the outgoing currents. Applying KCL to Circuit B of Figure 4-11 gives you the following:

$$i_s = i_R + i_1 + i_2$$

Because Circuit B shows devices connected in parallel, the voltage v is the same across each resistor. Using Ohm's law for each resistor and using the definition of conductance G (found in Chapter 3) yields the following expressions for the currents:

$$i_R = \frac{v}{R} = Gv \qquad i_1 = \frac{v}{R_1} = G_1 v \qquad i_2 = \frac{v}{R_2} = G_2 v$$

where $G = 1/R$, $G_1 = 1/R_1$, and $G_2 = 1/R_2$.

Substituting the values of i_R, i_1, and i_2 into $i_s = i_R + i_1 + i_2$ and factoring out the voltage v gives you

$$i_s = v(G + G_1 + G_2)$$

When you divide $i_1 = vG_1$ by $i_s = v(G + G_1 + G_2)$, you wind up with the following form of the desired current divider equation:

$$\frac{i_1}{i_s} = \frac{vG_1}{v(G + G_1 + G_2)}$$

$$\frac{i_1}{i_s} = \frac{G_1}{G + G_1 + G_2}$$

This form of the current divider equation is often referred to as a *current transfer function,* and it relates the ratio output current (current i_1 for this example) to the input source current (i_s in this example). With this equation, you can find the output current going through any device for a given input source current.

Algebraically solving for i_1 yields the following form of the current divider equation:

$$i_1 = i_s \left(\frac{G_1}{G + G_1 + G_2} \right)$$

You can find similar relationships for i_2 and i_R:

$$i_2 = i_s \left(\frac{G_2}{G + G_1 + G_2} \right)$$

$$i_R = i_s \left(\frac{G}{G + G_1 + G_2} \right)$$

These equations show that to find the current through a desired conductance, you simply multiply the input source current by the desired conductance divided by the total conductance of the parallel combination in the circuit. Thus, in Figure 4-11, the current through each resistor depends on the ratio of resistors multiplied by the input source current.

The current through each resistor is always less than and proportional to the supplied independent source current because the ratio of conductance is always less than 1. That idea offers a neat way to check your answer: For a parallel combination of resistors, the largest current goes through the smallest resistor because it has the least resistance, and the smallest current goes through the largest resistor.

Figuring out currents for parallel circuits

The current divider method provides a shortcut in finding the current through each device when all the devices are connected in parallel.

Try using current divider techniques to calculate the current through each resistor in Circuit A of Figure 4-12. Circuit A is a complex circuit with a voltage source, so first transform it into an equivalent circuit that has a current source and all the resistors connected in parallel, as in Circuit B.

Start by finding the source current in Circuit B. The transformation must be constrained by $v_s = i_s R$, where V = 24 volts and v_s = 6 Ω, so the source current for Circuit B, i_s, is 4 amps.

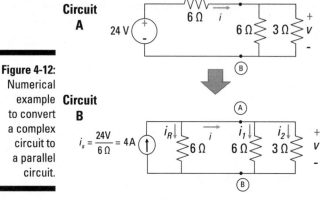

Illustration by Wiley, Composition Services Graphics

Figure 4-12:
Numerical example Circuit to convert a complex circuit to a parallel circuit.

After you know the source current, i_s, you can use the current divider equation for each resistor. Using the current divider method yields the following (see the preceding section for the derivation of this equation):

$$i_1 = 4 \text{ A} \cdot \left(\frac{\frac{1}{6 \, \Omega}}{\frac{1}{6 \, \Omega} + \frac{1}{6 \, \Omega} + \frac{1}{3 \, \Omega}} \right) = 1 \text{ A}$$

$$i_2 = 4 \text{ A} \cdot \left(\frac{\frac{1}{3 \, \Omega}}{\frac{1}{6 \, \Omega} + \frac{1}{6 \, \Omega} + \frac{1}{3 \, \Omega}} \right) = 2 \text{ A}$$

$$i_R = 4 \text{ A} \cdot \left(\frac{\frac{1}{6 \, \Omega}}{\frac{1}{6 \, \Omega} + \frac{1}{6 \, \Omega} + \frac{1}{3 \, \Omega}} \right) = 1 \text{ A}$$

These results show how easy it is to use the current divider method after you've transformed the circuit into a parallel circuit driven by a practical current source.

Finding currents when you have multiple voltage sources

Circuit A of Figure 4-13 has multiple voltage sources. What is i_1, the current through R_1? You can find the answer by transforming the circuit and using the current divider technique.

Transforming this circuit involves two stages. The first is converting the voltage sources to current sources and connecting all resistors in parallel. The second is combining the two current sources. You can then apply the current divider technique to find i_1.

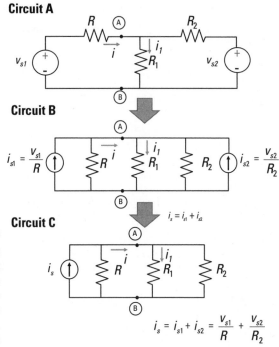

Illustration by Wiley, Composition Services Graphics

Figure 4-13:
Circuit
analysis
with mul-
tiple voltage
sources.

Circuit B shows the first part of the transformation of Circuit A: switching from voltage to current sources. For the transformed circuit to be equivalent to the original, the following equations have to hold true (see the earlier section "Converting to a parallel circuit with a current source" for these constraint equations):

$$i_{s1} = \frac{v_{s1}}{R}$$

$$i_{s2} = \frac{v_{s2}}{R_2}$$

Circuit C of Figure 4-13 completes the transformation. The two current sources are connected in parallel and combined to form one equivalent current source. The current sources point in the same direction, so you can add them up to get the following:

$$i_s = i_{s1} + i_{s2}$$

Figure 4-14 provides a numerical example for the circuit shown in Figure 4-13. Start by switching to current sources.

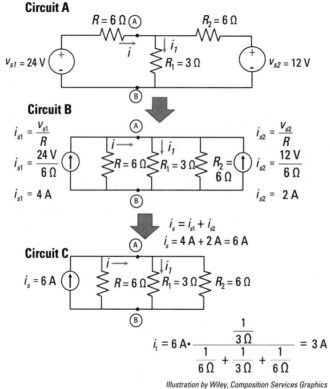

Figure 4-14: Numerical example of a circuit with multiple voltage sources.

The transformation of the two voltage sources to two current sources yields

$$i_{s1} = \frac{v_{s1}}{R} = \frac{24\text{ V}}{6\text{ }\Omega} = 4\text{ A}$$

$$i_{s2} = \frac{v_{s2}}{R_2} = \frac{12\text{ V}}{6\text{ }\Omega} = 2\text{ A}$$

You can see the results of the source transformations and equivalent resistance in Circuit B of Figure 4-14.

The current sources are connected in parallel and point in the same direction, so add them together:

$$i_s = i_{s1} + i_{s2} = 4\,A + 2\,A = 6\,A$$

Circuit C of Figure 4-14 shows the completed and simplified transformation that you use to calculate i_1. Use the current divider technique to find the current through the 3-Ω resistor:

$$i_1 = 6\,A \cdot \frac{\dfrac{1}{3\,\Omega}}{\dfrac{1}{6\,\Omega} + \dfrac{1}{3\,\Omega} + \dfrac{1}{6\,\Omega}} = 3\,A$$

Using the current divider technique repeatedly

When you see parts of a circuit with resistor combinations connected in parallel within other combinations of devices connected in parallel, you can use current divider techniques repeatedly. To see how this works, consider Circuit A in Figure 4-15. You can use current divider shortcuts repeatedly to find the current i_x through the 8-kΩ resistor.

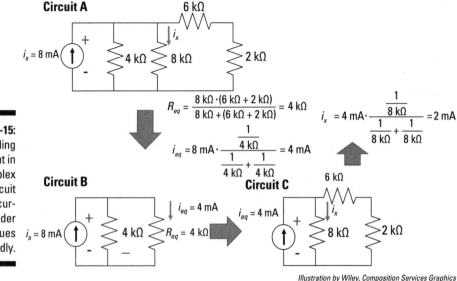

Figure 4-15: Finding current in a complex circuit using current divider techniques repeatedly.

Illustration by Wiley, Composition Services Graphics

The circuit includes a resistor series combination of 6 kΩ and 2 kΩ connected in parallel with the 8-kΩ resistor. For this resistor combination, total resistance yields

$$R_{eq} = \frac{8 \text{ k}\Omega \cdot (6 \text{ k}\Omega + 2 \text{ k}\Omega)}{8 \text{ k}\Omega + (6 \text{ k}\Omega + 2 \text{ k}\Omega)} = 4 \text{ k}\Omega$$

Circuit B shows the equivalent resistance. Calculate the current through R_{eq} using the current divider equation:

$$i_{eq} = 8 \text{ mA} \cdot \frac{\dfrac{1}{4 \text{ k}\Omega}}{\dfrac{1}{4 \text{ k}\Omega} + \dfrac{1}{4 \text{ k}\Omega}} = 4 \text{ mA}$$

However, the 4-milliamp current is split between the 4-kΩ resistor and the series resistor combination of 6 kΩ and 2 kΩ, as in Circuit C of Figure 4-15. Use the current divider method once again to find the current through the 8-kΩ resistor:

$$i_x = 4 \text{ mA} \cdot \frac{\dfrac{1}{8 \text{ k}\Omega}}{\dfrac{1}{8 \text{ k}\Omega} + \dfrac{1}{6 \text{ k}\Omega + 2 \text{ k}\Omega}} = 2 \text{ mA}$$

Even though the current divider shortcut for parallel circuits makes it so you don't have to use Kirchhoff's laws and Ohm's laws to find an unknown current, this technique was developed from the foundational equations of Kirchhoff's laws and Ohm's law.

Part II

Applying Analytical Methods for Complex Circuits

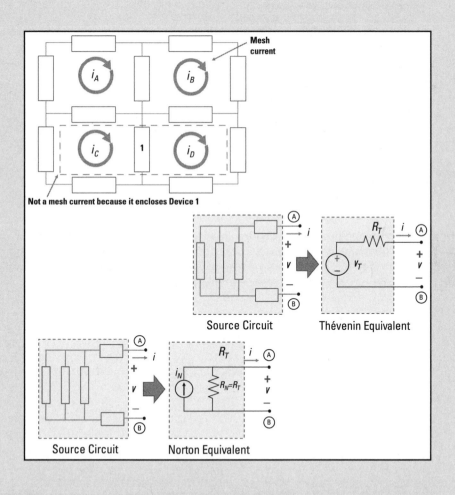

In this part . . .

- ✔ Practice node-voltage analysis in order to describe the voltages across each device in a circuit.

- ✔ Apply mesh-current analysis to circuits that have many devices connected in series.

- ✔ Deal with multiple current and voltage sources with the superposition technique.

- ✔ Simplify source circuits with the Thévenin and Norton theorems.

Chapter 5

Giving the Nod to Node-Voltage Analysis

In This Chapter

▶ Describing node-voltage analysis

▶ Applying Kirchhoff's current law to node-voltage analysis

▶ Putting node-voltage equations in matrix form

*Y*ou can describe voltages across each device in a circuit by using node-voltage analysis (NVA), one of the major techniques in circuit analysis. Better yet, NVA reduces the number of equations you have to deal with. I tell you all about the key ingredients of NVA — node voltages and reference nodes — in this chapter. I also walk you through the technique, first with a basic example and then with more-complex ones.

Getting Acquainted with Node Voltages and Reference Nodes

A *node* is a particular junction or point on a circuit. To use node voltages, you need to select a reference point (or *ground point*) defined as 0 volts. *Node voltages* are voltages at circuit nodes measured with respect to that reference node.

Figure 5-1 shows you the notation for node voltage variables as well as the voltage across each device. The voltages V_A and V_B are the node voltages measured with respect to a reference node, which is identified by the ground symbol. These node voltages can describe the voltages v_1, v_2, and v_3 for the three devices in the circuit.

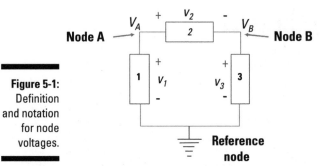

You calculate device voltage as the difference between two node voltages. Take the node voltage at the device's positive terminal minus the node voltage at the device's negative terminal.

Look at Figure 5-1. Because it takes two points to define a voltage, the device voltage v_1 is the difference between the node voltage V_A and the voltage of the reference node, 0 volts. Device 1 has its positive terminal connected to Node A and its negative terminal connected to the reference node, so here's the device voltage:

$$v_1 = V_A - 0 = V_A$$

Device 2 has its positive terminal connected to Node A and its negative terminal connected to Node B. Device 2's voltage v_2 is

$$v_2 = V_A - V_B$$

Device 3 has its positive terminal connected to Node B and its negative terminal connected to the reference node. Because Device 3 is connected to a reference node, its voltage v_3 is

$$v_3 = V_B - 0 = V_B$$

Testing the Waters with Node-Voltage Analysis

With node-voltage analysis, or NVA, the goal is to find the voltages across the devices in a circuit. You first apply Kirchhoff's current law (KCL), which states that the sum of incoming currents is equal to the sum of the outgoing currents at any node in the circuit. (See Chapter 3 for more on KCL.) With KCL, you can find a set of equations to determine the unknown node voltages. And when you know all the node voltages in the circuit, you can find the voltages across each device in terms of the node voltages.

In other words, node-voltage analysis involves the following steps:

1. **Select a reference (ground) node.**

   The reference node doesn't have to be actually connected to ground. You simply identify the node that way for the analysis.

   Because a reference node has 0 volts, you can simplify the analysis by choosing a node where a large number of devices are connected as your reference node.

2. **Formulate a KCL equation for each nonreference node.**

3. **Express the device currents in terms of node voltages by using device relationships such as Ohm's law.**

4. **Substitute the device equations from Step 3 into the KCL equations of Step 2.**

   Simplify the equations to put them in standard form.

5. **Solve the system of equations to find the unknown node voltages.**

   Rearrange the standard-form equations into matrix form and use matrix software to solve for the node voltages (or solve very simple systems of equations using other techniques from linear algebra).

Because Step 1 is easy, the next sections walk you through the rest of the steps of node-voltage analysis.

What goes in must come out: Starting with KCL at the nodes

After you choose a reference node, the first step in finding node voltage equations is to set up the Kirchhoff's current law (KCL) equations for a given circuit. I use the circuit in Figure 5-2 to show you how to develop these equations. The ground symbol at the bottom of the figure tells you which node is the reference node (the node having 0 volts).

This circuit has two node voltages, V_A and V_B, and four element currents, i_s, i_1, i_2, and i_3.

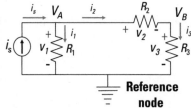

Figure 5-2:
A circuit with a reference node and two node voltages.

Illustration by Wiley, Composition Services Graphics

At Node A, the source current i_s splits into i_1 and i_2. Here's the KCL equation for the device currents at Node A:

$$\text{in} = \text{out} \quad \rightarrow \quad i_s = i_1 + i_2$$

And here's the KCL equation for Node B:

$$\text{in} = \text{out} \quad \rightarrow \quad i_2 = i_3$$

So now you know about the currents at play in Figure 5-2. How do you get the voltages? By applying Ohm's law, as I explain next.

Describing device currents in terms of node voltages with Ohm's law

Ohm's law expresses a linear relationship between voltage and current when the device in question is a resistor. You need Ohm's law to describe a device's current in terms of its node voltages. First, you determine what the node voltages are and find the device voltages. Then you substitute the node voltage expressions of the device currents into KCL and get the set of node voltage equations to be solved.

Look at the node voltages on either side of resistor R_1 in Figure 5-2. The device voltage is the difference in node voltages. Because the negative terminal is connected to a reference node, the voltage v_1 across resistor R_1 is

$$v_1 = V_A - 0$$
$$v_1 = V_A$$

The voltage v_2 for Device 2 is the difference between the node voltages at Nodes A and B. The device's positive terminal is connected to Node A, and its negative terminal is connected to Node B, so

$$v_2 = V_A - V_B$$

The negative terminal for Device 3 is connected to a reference node, so the voltage v_3 is simply

$$v_3 = V_B - 0$$
$$v_3 = V_B$$

Now apply Ohm's law ($i = v/R$) to express the device currents through R_1, R_2, and R_3 in terms of the node voltages. Using Ohm's law produces the following device currents:

$$i_1 = \frac{v_1}{R_1} = \left(\frac{1}{R_1}\right)V_A$$

$$i_2 = \frac{v_2}{R_2} = \left(\frac{1}{R_2}\right)(V_A - V_B)$$

$$i_3 = \frac{v_3}{R_3} = \left(\frac{1}{R_3}\right)V_B$$

You can now substitute these device-current expressions into the KCL equations at Nodes A and B (see the preceding section for the KCL equations). You wind up with:

Node A: $i_s = i_1 + i_2$ $\rightarrow$ $i_s = \left(\frac{1}{R_1}\right)V_A + \left(\frac{1}{R_2}\right)(V_A - V_B)$

Node B: $i_2 = i_3$ $\rightarrow$ $\left(\frac{1}{R_2}\right)(V_A - V_B) = \left(\frac{1}{R_3}\right)V_B$

Collect like terms and rearrange these two node equations to get the following:

Node A: $\left(\dfrac{1}{R_1}+\dfrac{1}{R_2}\right)V_A-\left(\dfrac{1}{R_2}\right)V_B=i_s$

Node B: $-\left(\dfrac{1}{R_2}\right)V_A+\left(\dfrac{1}{R_2}+\dfrac{1}{R_3}\right)V_B=0$

These two node voltage equations are said to be in standard form, but you can easily put this set of equations into matrix form.

Putting a system of node voltage equations in matrix form

The node voltage equations (see the preceding section) give you a system of linear equations, which you can solve using matrices. Of course, you can skip the matrices if the system is simple and you want to use other techniques from linear algebra, such as back substitution, to find the answers. But in most cases, using matrices is faster and easier, especially if you have a large and complicated circuit.

Here's how to transform node voltage equations from standard form to matrix form:

1. **Take the coefficients (of resistors or conductances) of the node voltages to form a square matrix.**

   Make sure the variable terms are in the same order in all your node voltage equations before setting up the matrix.

   A square matrix has the same number of columns and rows. Each column holds all the coefficients on a particular variable, and each row holds all the coefficients from a particular equation.

2. **Multiply the coefficient matrix from Step 1 by a column vector of the node voltages (the variables you want to solve for).**

   A column vector is a single-column matrix. The number of rows in the column vector should equal the number of columns in the square matrix.

   In the column vector, write the variables in the order in which they appear in your node voltage equations.

3. **Write the right side of each node equation as a vector element to form a column vector of current sources when combining the system of node equations.**

   The column vector of current sources should have the same number of rows as the column vector of node voltages.

   In this column vector, write the current sources that appear to the right of the equal signs in your node voltage equations.

When you translate the set of node voltage equations from the preceding section into matrix form, you wind up with the following description of the circuit:

$$
\begin{bmatrix} \dfrac{1}{R_1}+\dfrac{1}{R_2} & -\dfrac{1}{R_2} \\ -\dfrac{1}{R_2} & \dfrac{1}{R_2}+\dfrac{1}{R_3} \end{bmatrix} \begin{bmatrix} V_A \\ V_B \end{bmatrix} = \begin{bmatrix} i_s \\ 0 \end{bmatrix}
$$

This matrix equation follows the form of $Ax = b$, where A consists of a matrix of coefficients of resistors or conductances, x is a vector of unknown node voltages, and b is vector of independent current sources.

Confirming diagonal symmetry in the square matrix is a useful way to check that your node voltage equations are right. For circuits with independent sources, you should see positive values along one diagonal and negative values along the opposite diagonal (the off-diagonal elements).

Solving for unknown node voltages

After you have your system of node voltage equations in matrix form, you're ready to solve for the unknown node voltages. You could solve simple matrices for the node voltages using Cramer's rule or other techniques from linear algebra. But for circuits with a large number of elements, use matrix software or a graphing calculator. For instance, you can find node voltages by multiplying the inverse of the coefficient matrix by the answer matrix (the column vector of current sources) on your graphing calculator: $A^{-1}b = x$.

Matrix software is great for doing calculations, but it doesn't develop the node voltage equations for you. Make sure you know how to set up the matrix problem to help you solve for the node voltages. Fortunately, some circuit analysis software does solve for the unknown voltages. You need to build the circuit graphically (depending on the software), and the software performs the required calculations.

Applying the NVA Technique

If you've reviewed the earlier sections in this chapter, then you're ready to set up some node voltage equations with numerical examples. When you have a voltage source with one of its terminals connected to a reference node, the node voltage is simply equal to the voltage source. Although doing so requires a little more work, if you're comfortable with current sources, you can always transform a voltage source into a current source. (If you're wondering what *NVA* and *node voltage equations* are, spend some time with the first part of this chapter before moving on.)

Solving for unknown node voltages with a current source

Formulating the node voltage equations leads to a linear system of equations. You can see what I mean by working through the NVA process I outline in the earlier section "Testing the Waters with Node-Voltage Analysis." Try finding the voltages and currents for the devices in the circuit in Figure 5-3. (Note that Figure 5-3 is the same as Figure 5-2 but with numbers given for R_1, R_2, and R_3.)

Figure 5-3:
Numerical example of node-voltage analysis.

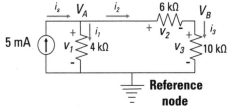

Illustration by Wiley, Composition Services Graphics

Start by identifying your reference node. I marked my chosen reference node with the ground symbol in Figure 5-3. Now you can form the KCL equations for Nodes A and B:

Node A: in = out $\rightarrow$ $i_s = i_1 + i_2$

Node B: in = out $\rightarrow$ $i_2 = i_3$

Next, express the device currents in terms of node voltages by using Ohm's law (see the earlier section "Getting Acquainted with Node Voltages and Reference Nodes" for help writing the device currents). You wind up with the following equations:

$$i_1 = \frac{v_1}{R_1} = \left(\frac{1}{4\ \text{k}\Omega}\right)V_A$$

$$i_2 = \frac{v_2}{R_2} = \left(\frac{1}{6\ \text{k}\Omega}\right)(V_A - V_B)$$

$$i_3 = \frac{v_3}{R_3} = \left(\frac{1}{10\ \text{k}\Omega}\right)V_B$$

Go ahead and substitute these current values into the KCL equations. Then rearrange the equations to put them in standard form. Here's the equation for Node A:

$$i_1 + i_2 = i_s$$

$$\left(\frac{1}{4\ \text{k}\Omega}\right)V_A + \left(\frac{1}{6\ \text{k}\Omega}\right)(V_A - V_B) = 5\ \text{mA}$$

$$\left(\frac{1}{4\ \text{k}\Omega} + \frac{1}{6\ \text{k}\Omega}\right)V_A - \left(\frac{1}{6\ \text{k}\Omega}\right)V_B = 5\ \text{mA}$$

And here's the equation for Node B:

$$i_2 = i_3$$

$$\left(\frac{1}{6\ \text{k}\Omega}\right)(V_A - V_B) = \left(\frac{1}{10\ \text{k}\Omega}\right)V_B$$

$$-\left(\frac{1}{6\ \text{k}\Omega}\right)V_A + \left(\frac{1}{6\ \text{k}\Omega} + \frac{1}{10\ \text{k}\Omega}\right)V_B = 0$$

You now have a system of linear equations for Node A and Node B — two equations with two variables. Write this system of equations in matrix form (for details, see the earlier section "Putting a system of node voltage equations in matrix form"). The resulting matrix looks like this:

$$\begin{bmatrix} \dfrac{1}{4\ \text{k}\Omega} + \dfrac{1}{6\ \text{k}\Omega} & -\dfrac{1}{6\ \text{k}\Omega} \\ -\dfrac{1}{6\ \text{k}\Omega} & \dfrac{1}{6\ \text{k}\Omega} + \dfrac{1}{10\ \text{k}\Omega} \end{bmatrix} \begin{bmatrix} V_A \\ V_B \end{bmatrix} = \begin{bmatrix} 5\ \text{mA} \\ 0 \end{bmatrix}$$

Use your calculator or matrix software to solve for V_A and V_B. You wind up with the following node voltages:

$$\begin{bmatrix} V_A \\ V_B \end{bmatrix} = \begin{bmatrix} 16\text{ V} \\ 10\text{ V} \end{bmatrix}$$

Now calculate the device voltages by finding the difference between two node voltages (see the first section in this chapter for details). Given these node voltages, the voltages across the resistors are

$$v_1 = V_A - 0 = 16\text{ V}$$

$$v_2 = V_A - V_B = 16\text{ V} - 10\text{ V} = 6\text{ V}$$

$$v_3 = V_B - 0 = 10\text{ V}$$

To complete the analysis, use Ohm's law to calculate the current for each resistor:

$$i_1 = \frac{v_1}{R_1} = \frac{16\text{ V}}{4\text{ k}\Omega} = 4\text{ mA}$$

$$i_2 = \frac{v_2}{R_2} = \frac{6\text{ V}}{6\text{ k}\Omega} = 1\text{ mA}$$

$$i_3 = \frac{v_3}{R_3} = \frac{10\text{ V}}{10\text{ k}\Omega} = 1\text{ mA}$$

After you finish the calculations, check whether your answers make sense. You can verify your results by applying Kirchhoff's and Ohm's laws. You can also verify that devices connected in series have the same current and that devices connected in parallel have the same voltage.

The answers make sense for this problem because the outgoing currents i_1 and i_2 at Node A add up to 5 milliamps. Also, resistor R_1, with a resistance of 4 kΩ, has four times the current of the series combination of R_2 and R_3, whose resistance totals 16 kΩ. And R_2 and R_3 have the same current, as is true for any series combination.

Dealing with three or more node equations

The node voltage approach is most useful when the circuit has three or more node voltages. You can use the same step-by-step process you use for circuits with two nodes, which I show you earlier in "Testing the Waters with Node-Voltage Analysis."

The circuit in Figure 5-4 has four nodes: A, B, C, and a reference node of 0 volts marked with a ground symbol. You want to find the voltage across each device in the circuit.

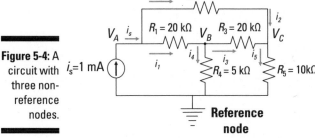

Figure 5-4: A circuit with $i_s=1$ mA three non-reference nodes.

Illustration by Wiley, Composition Services Graphics

At Node A, you have incoming current i_s and outgoing currents i_1 and i_2. At Node B, you have incoming current i_1 and outgoing currents i_3 and i_4. And at Node C, you have incoming currents i_2 and i_3 and outgoing current i_5. By applying the KCL equations at Nodes A, B, and C, you wind up with the following:

Node A :   in = out : $i_s = i_1 + i_2$

Node B :   in = out : $i_1 = i_3 + i_4$

Node C :   in = out : $i_2 + i_3 = i_5$

Next, express the device currents in terms of node voltages using Ohm's law (see the earlier section "Getting Acquainted with Node Voltages and Reference Nodes" for info on writing the device currents):

$$i_1 = \frac{v_1}{R_1} = \left(\frac{1}{20 \text{ k}\Omega}\right)(V_A - V_B)$$

$$i_2 = \frac{v_2}{R_2} = \left(\frac{1}{5 \text{ k}\Omega}\right)(V_A - V_C)$$

$$i_3 = \frac{v_3}{R_3} = \left(\frac{1}{20 \text{ k}\Omega}\right)(V_B - V_C)$$

$$i_4 = \frac{v_4}{R_4} = \left(\frac{1}{5 \text{ k}\Omega}\right)V_B$$

$$i_5 = \frac{v_5}{R_5} = \left(\frac{1}{10 \text{ k}\Omega}\right)V_C$$

Substitute these device-current equations into the KCL equations. Then algebraically rearrange the equations to put them in standard form. Here's the equation for Node A:

$$i_1 + i_2 = i_s$$

$$\left(\frac{1}{20\text{ k}\Omega}\right)(V_A - V_B) + \left(\frac{1}{5\text{ k}\Omega}\right)(V_A - V_C) = 1\text{ mA}$$

$$\left(\frac{1}{5\text{ k}\Omega} + \frac{1}{20\text{ k}\Omega}\right)V_A - \left(\frac{1}{20\text{ k}\Omega}\right)V_B - \left(\frac{1}{5\text{ k}\Omega}\right)V_C = 1\text{ mA}$$

Here's the equation for Node B:

$$i_3 + i_4 = i_1$$

$$\left(\frac{1}{20\text{ k}\Omega}\right)(V_B - V_C) + \left(\frac{1}{5\text{ k}\Omega}\right)V_B = \left(\frac{1}{20\text{ k}\Omega}\right)(V_A - V_B)$$

$$\left(-\frac{1}{20\text{ k}\Omega}\right)V_A + \left(\frac{1}{20\text{ k}\Omega} + \frac{1}{20\text{ k}\Omega} + \frac{1}{5\text{ k}\Omega}\right)V_B - \left(\frac{1}{20\text{ k}\Omega}\right)V_C = 0$$

And here's the equation for Node C:

$$i_2 + i_3 = i_5$$

$$\left(\frac{1}{5\text{ k}\Omega}\right)(V_A - V_C) + \left(\frac{1}{20\text{ k}\Omega}\right)(V_B - V_C) = \left(\frac{1}{10\text{ k}\Omega}\right)V_C$$

$$\left(-\frac{1}{5\text{ k}\Omega}\right)V_A - \left(\frac{1}{20\text{ k}\Omega}\right)V_B + \left(\frac{1}{5\text{ k}\Omega} + \frac{1}{20\text{ k}\Omega} + \frac{1}{10\text{ k}\Omega}\right)V_C = 0$$

Simplifying the coefficients gives you the following set of node voltage equations for Nodes A, B, and C:

$$\left(\frac{1}{4\text{ k}\Omega}\right)V_A - \left(\frac{1}{20\text{ k}\Omega}\right)V_B - \left(\frac{1}{5\text{ k}\Omega}\right)V_C = 1\text{ mA}$$

$$\left(-\frac{1}{20\text{ k}\Omega}\right)V_A + \left(\frac{3}{10\text{ k}\Omega}\right)V_B - \left(\frac{1}{20\text{ k}\Omega}\right)V_C = 0\text{ mA}$$

$$\left(-\frac{1}{5\text{ k}\Omega}\right)V_A - \left(\frac{1}{20\text{ k}\Omega}\right)V_B + \left(\frac{7}{20\text{ k}\Omega}\right)V_C = 0\text{ mA}$$

Now put this system of node voltage equations in matrix form. (I explain how to do this earlier in "Putting a system of node-voltage equations in matrix form.")

$$
\begin{bmatrix}
\dfrac{1}{4\ \text{k}\Omega} & -\dfrac{1}{20\ \text{k}\Omega} & -\dfrac{1}{5\ \text{k}\Omega} \\[2mm]
-\dfrac{1}{20\ \text{k}\Omega} & \dfrac{3}{10\ \text{k}\Omega} & -\dfrac{1}{20\ \text{k}\Omega} \\[2mm]
-\dfrac{1}{5\ \text{k}\Omega} & -\dfrac{1}{20\ \text{k}\Omega} & \dfrac{7}{20\ \text{k}\Omega}
\end{bmatrix}
\begin{bmatrix}
V_A \\ V_B \\ V_C
\end{bmatrix}
=
\begin{bmatrix}
1\ \text{mA} \\ 0\ \text{mA} \\ 0\ \text{mA}
\end{bmatrix}
$$

The preceding matrix equation is of the form $Ax = b$. Notice that the square matrix is symmetrical along the diagonal the diagonal terms are positive, and the off-diagonal terms are negative, all of which suggests that you've converted to matrix form correctly.

Plug the matrix equation into your calculator or matrix software, giving you

$$
\begin{bmatrix}
V_A \\ V_B \\ V_C
\end{bmatrix}
=
\begin{bmatrix}
8.723\ \text{V} \\ 2.340\ \text{V} \\ 5.319\ \text{V}
\end{bmatrix}
$$

Now that you know the voltages for Nodes A, B, and C, you can determine the voltages across Devices 1 through 5:

$$v_1 = V_A - V_B = 6.383\ \text{V}$$
$$v_2 = V_A - V_C = 3.404\ \text{V}$$
$$v_3 = V_B - V_C = -2.979\ \text{V}$$
$$v_4 = V_B - 0 = 2.3404\ \text{V}$$
$$v_5 = V_C - 0 = 5.319\ \text{V}$$

To complete the analysis, find the current through each device:

$$i_1 = \frac{v_1}{R_1} = \frac{6.383\ \text{V}}{20\ \text{k}\Omega} = 0.3192\ \text{mA}$$

$$i_2 = \frac{v_2}{R_2} = \frac{3.404\ \text{V}}{5\ \text{k}\Omega} = 0.6807\ \text{mA}$$

$$i_3 = \frac{v_4}{R_4} = \frac{-2.979\ \text{V}}{20\ \text{k}\Omega} = -0.1490\ \text{mA}$$

$$i_4 = \frac{v_4}{R_4} = \frac{2.3404\ \text{V}}{5\ \text{k}\Omega} = 0.4681\ \text{mA}$$

$$i_5 = \frac{v_5}{R_5} = \frac{5.319\ \text{V}}{10\ \text{k}\Omega} = 0.5319\ \text{mA}$$

These results make sense because they satisfy the KCL equations at each of the three nodes.

Working with Voltage Sources in Node-Voltage Analysis

REMEMBER

When a voltage source is connected to a node, you end up with fewer unknown node voltage equations because one of the node voltages is given in terms of the known voltage source. Here's how the node voltages compare if you have a voltage source:

✔ If the negative terminal of the voltage source is connected to a reference node, then the voltage of the node connected to the positive terminal of the voltage source has to be equal to the source voltage.

✔ If the voltage source terminals are connected to two nonreference nodes, then the difference between the two node voltages is simply the source voltage. So if you know one node voltage, you get the other by adding or subtracting the source voltage to or from the known node voltage.

If you're more comfortable dealing with current sources, you can perform a source transformation by replacing the voltage source and resistors connected in series with an equivalent current source and resistors connected in parallel. I show you how to transform independent sources in Chapter 4.

Figure 5-5 shows that the negative terminal of a voltage source is usually given as 0 volts. As you can see, Circuit A has two voltage sources and three nonzero nodes. Through source transformation, you can transform the circuit into Circuit B, which has only one nonreference node.

Figure 5-5:
Using
source
transformation of
voltage
sources for
NVA.

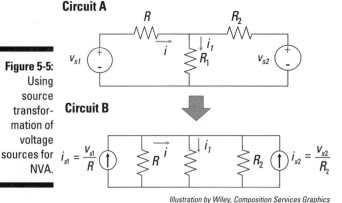

Illustration by Wiley, Composition Services Graphics

When you apply node-voltage analysis in Circuit B, you wind up with the following equation:

$$\left(\frac{1}{R_1}+\frac{1}{R_2}+\frac{1}{R_3}\right)V_A=\frac{v_{s1}}{R_1}+\frac{v_{s2}}{R_2}$$

You get the same result using source transformation by noting that $V_A = V_{s1}$ and $V_C = v_{s2}$. The next example illustrates the technique by relating the node voltages to the voltage source.

Sometimes you encounter circuits with two voltage sources that don't have a common node. One voltage source is connected to a reference node, and the other voltage source has terminals connected to nonreference nodes, as in Figure 5-6.

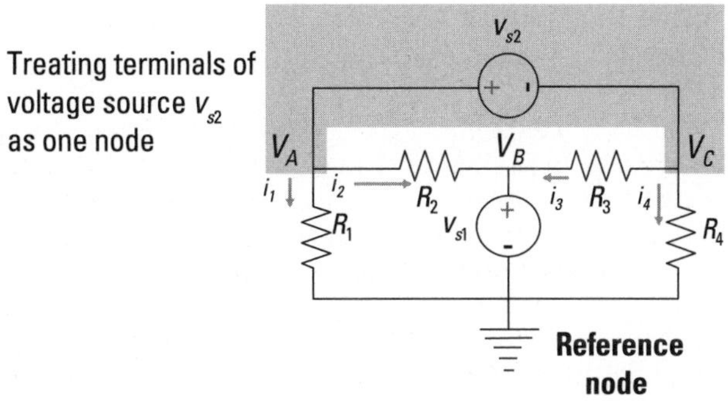

Treating terminals of voltage source v_{s2} as one node

Figure 5-6:
Dealing with ungrounded voltage source for NVA.

Consider the voltage source at the top of Figure 5-6. Currents i_1 through i_4 leave and enter through the negative and positive terminals of v_{s2}, which leads to the following KCL equation:

$$i_1 + i_2 + i_3 + i_4 = 0$$

You can express these node voltages in the KCL equation in the following expression:

$$\frac{V_A}{R_1}+\frac{V_A-V_B}{R_2}+\frac{V_C-V_B}{R_3}+\frac{V_C}{R_4}=0$$

The source voltage v_{s1} at Node B is connected to a reference node, which means that

$$V_B = v_{s1}$$

Because v_{s2} is connected at Nodes A and C, the voltage across v_{s2} is the difference between the node voltages at these nodes:

$$V_A - V_C = v_{s2}$$
$$V_C = V_A - v_{s2}$$

Replace V_B and V_C in the KCL equation to get the following expression:

$$\frac{V_A}{R_1} + \frac{V_A - v_{s1}}{R_2} + \frac{V_A - v_{s2} - v_{s1}}{R_3} + \frac{V_A - v_{s2}}{R_4} = 0$$

Put the source voltages on one side of the equation, which gives you

$$\left(\frac{1}{R_1} + \frac{1}{R_2} + \frac{1}{R_3} + \frac{1}{R_4}\right)V_A = \left(\frac{1}{R_2} + \frac{1}{R_3}\right)v_{s1} + \left(\frac{1}{R_3} + \frac{1}{R_4}\right)v_{s2}$$

This equation now has one node voltage term.

Now, suppose the desired output voltage is the voltage across resistor R_4, connected to V_C. Substitute $V_C + v_{s2}$ for V_A into the preceding equation:

$$\left(\frac{1}{R_1} + \frac{1}{R_2} + \frac{1}{R_3} + \frac{1}{R_4}\right)(V_C + v_{s2}) = \left(\frac{1}{R_2} + \frac{1}{R_3}\right)v_{s1} + \left(\frac{1}{R_3} + \frac{1}{R_4}\right)v_{s2}$$

Now simplify the equation:

$$\left(\frac{1}{R_1} + \frac{1}{R_2} + \frac{1}{R_3} + \frac{1}{R_4}\right)V_C = \left(\frac{1}{R_2} + \frac{1}{R_3}\right)v_{s1} - \left(\frac{1}{R_1} + \frac{1}{R_2}\right)v_{s2}$$

Now you have one equation with the node voltage V_C. This equation is easily solvable using algebra after you plug in some numbers for the resistors and voltage sources.

Chapter 6

Getting in the Loop on Mesh Current Equations

. . . .

In This Chapter

▶ Describing mesh currents

▶ Applying Kirchhoff's voltage laws (KVL) to mesh-current analysis

▶ Analyzing a couple of circuits

. . . .

Mesh-current analysis (also known as *loop-current analysis*) can help reduce the number of equations you need to solve simultaneously when dealing with circuits that have many devices connected in multiple loops. This method is nothing but Kirchhoff's voltage law adapted for circuits with unique configurations.

In this chapter, I explain how to recognize meshes and assign mesh currents in order to calculate device currents and voltages.

Windowpanes: Looking at Meshes and Mesh Currents

To understand how mesh-current analysis works its magic, you need to know what a mesh is. Meshes occur in *planar circuits* — circuits that are drawn in a single plane or flat surface, without crossovers. The single plane is divided into a number of distinct areas, each of which looks like a windowpane, and the boundary of each windowpane is called a *mesh* of the circuit. The mesh can't enclose any devices — devices must fall on the boundary of the loop.

A *mesh current* is the current flowing around a mesh of the circuit. You get to choose the direction of the mesh current for your analysis. If the answer comes out negative, then the actual current is opposite the mesh current.

To help you distinguish between mesh currents and nonmesh currents, check out the planar circuit in Figure 6-1, which shows the notation of mesh current variables. The currents i_A, i_B, i_C, and i_D are all mesh currents, but the dashed rectangular box is not a mesh current because it encloses Device 1.

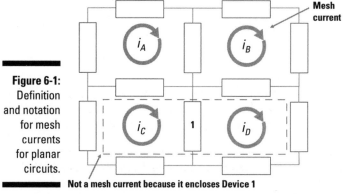

Figure 6-1:
Definition
and notation
for mesh
currents
for planar
circuits.

Not a mesh current because it encloses Device 1

Illustration by Wiley, Composition Services Graphics

Relating Device Currents to Mesh Currents

So why should you care about mesh currents in the first place? Because using mesh currents to describe the currents flowing through each device in a circuit reduces the number of equations you need to solve simultaneously. To see the relationship between mesh currents and the device currents, consider Figure 6-2. In this figure, the mesh currents are i_A, i_B, and i_C. Currents i_1 through i_9 are the device currents.

You get to choose the direction of the mesh currents. I recommend making them all point in the same clockwise (or counterclockwise) direction so that it's easier to formulate the equations consistently. With a little practice, you'll be able to write the mesh equations simply by looking at the circuit diagram.

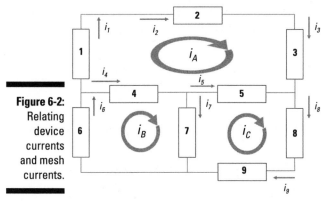

Illustration by Wiley, Composition Services Graphics

Figure 6-2: Relating device currents and mesh currents.

After choosing a current direction for each mesh, you can describe the currents for all the devices in the circuit. Here are the key points in describing device currents in terms of mesh currents:

✔ **If a device has only one mesh current flowing through it and the device current flows in the same direction as the mesh current, the currents are equal.** For example, because i_1, i_2, and i_3 flow in the same direction as mesh current i_A, you can express the device currents i_1, i_2, and i_3 as follows:

$$i_1 = i_2 = i_3 = i_A$$

You have similar situations for Devices 6, 8, and 9, where the device currents likewise flow in the same direction as their respective mesh currents:

$$i_6 = i_B$$

$$i_8 = i_9 = i_C$$

If, however, the device current flows in the opposite direction of the mesh current, you put a negative sign on the mesh current.

✔ **If a device has two mesh currents flowing through it, the device current equals the algebraic sum of the mesh currents.** Remember that a mesh current is positive if it's in the same direction as the device current and negative if it's in the opposite direction.

Consider Device 4, which has mesh currents i_A and i_B flowing through it. To describe i_4, find the sum of the mesh currents, making i_A negative because it flows in the opposite direction of i_4. Mathematically, you describe this device current as

$$i_4 = -i_A + i_B$$

Devices 5 and 7 similarly have multiple mesh currents flowing through each device, so you get the following equations:

$$i_5 = -i_A + i_C$$

$$i_7 = i_B - i_C$$

Generating the Mesh Current Equations

Mesh-current analysis is straightforward when planar circuits have voltage sources because you can easily develop Kirchhoff's voltage law (KVL) equations for each loop of the circuit. KVL says that the sum of the voltage rises and drops for any loop — or mesh — is equal to zero. (Check out Chapter 3 for more on KVL.) With KVL, Ohm's law, and a few substitutions, you can find device currents and voltages given only the source voltages and resistors.

Here's how it works: After choosing the direction of the current in each mesh, you look at the circuit diagram and describe the device currents in terms of the mesh currents. Substituting the device currents into Ohm's law gives you device voltages in terms of the mesh currents. And substituting the device voltages into the KVL equations gives you the source voltages in terms of the mesh currents. At that point, you have a system of linear equations, and you can solve for the mesh currents using matrices. With the mesh currents, you finish analyzing the circuit, finding the currents and voltages for all your devices.

Here's the step-by-step process:

1. **Select a current direction for each mesh.**

   See the preceding section for info on selecting the current direction.

2. **Formulate the KVL equations for each mesh.**

3. **Express the device voltages in terms of mesh currents using device relationships such as Ohm's law.**

   First use Ohm's law ($v = iR$) to relate device voltage to device current. Then replace the device current with its equivalent in terms of mesh current. I tell you how to express device current in terms of mesh currents in the preceding section.

4. **Substitute the device equations from Step 3 into the KVL equations from Step 2.**

5. **Put the equations in matrix form and solve the equations.**

   When you know the mesh currents, you can plug those values into the earlier equations to find the device currents and voltages.

The following sections walk you through Steps 2 through 5.

Finding the KVL equations first

To show you how to develop mesh current equations, I need a circuit with multiple loops. Enter Figure 6-3, which has two mesh currents (i_A and i_B) and five devices in the circuit. I decided to have both mesh currents flow clockwise.

KVL says that the sum of the voltage rises and drops for any loop is equal to zero. So for Mesh A, the KVL equation is

$$v_1 + v_2 = v_{s1}$$

For Mesh B, the KVL equation is

$$-v_2 + v_3 = -v_{s2}$$

Figure 6-3:
Demonstration of mesh-current analysis.

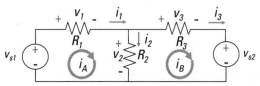

Illustration by Wiley, Composition Services Graphics

Ohm's law: Putting device voltages in terms of mesh currents

Ohm's law relates voltage and current. After you write device currents in terms of mesh currents, you can use Ohm's law to express the device voltages in terms of mesh currents.

First express the device currents in terms of the mesh currents (see the earlier section "Relating Device Currents to Mesh Currents" for details). In Figure 6-3, because the device current i_1 and the mesh current i_A point in the same direction through resistor R_1, the currents are equal:

$$i_1 = i_A$$

Now consider the current i_2 flowing through resistor R_2, which has two mesh currents flowing through it. In this case, i_A is in the same direction as i_2, but i_B is in the opposite direction of i_2, so i_B must be negative. You get the following expression for i_2:

$$i_2 = i_A - i_B$$

As for resistor R_3, you have its current i_3 equal to mesh current i_B:

$$i_3 = i_B$$

Now apply Ohm's law ($v = iR$) to relate current to voltage — and then do a little substitution. By replacing the device currents with their mesh current equivalents, you express the device voltages in terms of mesh currents. Here are the Ohm's law relationships for R_1, R_2, and R_3:

$$v_1 = i_1 R_1 = i_A R_1$$

$$v_2 = i_2 R_2 = (i_A - i_B) R_2$$

$$v_3 = i_3 R_3 = i_B R_3$$

Substituting the device voltages into the KVL equations

After you've expressed all the device currents in terms of the mesh currents, you're ready to substitute the device voltages v_1, v_2, and v_3 into the KVL equations for Meshes A and B. The result?

Mesh A: $\quad v_1 + v_2 = v_{s1} \quad \rightarrow \quad i_A R_1 + (i_A - i_B) R_2 = v_{s1}$

Mesh B: $\quad -v_2 + v_3 = -v_{s2} \quad \rightarrow \quad -(i_A - i_B) R_2 + i_B R_3 = -v_{s2}$

Then collect like terms and rearrange the preceding equations. Here are the equations for Meshes A and B in standard form:

Mesh A: $\quad (R_1 + R_2) i_A - R_2 i_B = v_{s1}$

Mesh B: $\quad -R_2 i_A + (R_2 + R_3) i_B = -v_{s2}$

Putting mesh current equations into matrix form

Together, the KVL equations for the meshes create a system of linear equations. The next step is to put the equations in matrix form so you can easily find the mesh currents using matrix software.

First, make sure your equations are in standard form (as they are in the preceding section). The standard form allows you to easily rearrange the mesh equations into matrix form: Resistors go in the coefficient matrix, mesh currents go in the variable vector, and source voltages go in the column vector of sources.

For Figure 6-3, the set of mesh equations in standard form (from the preceding section) becomes the following matrix form:

$$\begin{bmatrix} R_1 + R_2 & -R_2 \\ -R_2 & R_2 + R_3 \end{bmatrix} \begin{bmatrix} i_A \\ i_B \end{bmatrix} = \begin{bmatrix} v_{s1} \\ v_{s2} \end{bmatrix}$$

Notice the symmetry with respect to the main diagonal. The matrix has positive values along the main diagonal, and the off-diagonal terms have negative values.

 For circuits with independent sources, the matrix of resistors has a symmetry that can serve as a useful check when you're developing the mesh current equations. If you've set up the equations correctly, the off-diagonal terms will be symmetric with respect to the main diagonal. The terms along the main diagonal will be positive, and the off-diagonal terms will be negative or zero.

Solving for unknown currents and voltages

With the KVL equations in matrix form, you can solve for the mesh currents. When you know the mesh currents, the rest of the analysis is a snap. In deriving the KVL equations, you wrote some equations for the circuit devices that were in terms of the mesh currents (see the earlier section "Ohm's law: Putting device voltages in terms of mesh currents"). Go back to those equations, plug in the numbers, and do the math. I show you the whole process with some numbers in the next section.

Crunching Numbers: Using Meshes to Analyze Circuits

This section offers some numerical examples to show you how mesh-current analysis works. The first example involves two meshes, and the second example involves three.

Tackling two-mesh circuits

This section walks you through mesh-current analysis when you have two equations, one for Mesh A and one for Mesh B. In Figure 6-4, I decided to give both meshes a clockwise current. The next step is to apply KVL to Mesh A and B to arrive at the following mesh equations:

$$\text{Mesh A}: \quad v_{s1} = v_1 + v_2 \quad \rightarrow \quad v_1 + v_2 = 15 \text{ V}$$

$$\text{Mesh B}: \quad -v_2 + v_3 = -v_{s2} \quad \rightarrow \quad -v_2 + v_3 = -10 \text{ V}$$

Figure 6-4:
Mesh-current analysis with two meshes.

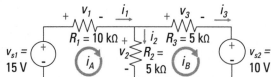

Illustration by Wiley, Composition Services Graphics

Next, write the device currents in terms of mesh currents. Then express the device currents in terms of the mesh currents using Ohm's law:

$$v_1 = i_1 R_1$$
$$= R_1 i_A = (10 \text{ k}\Omega) i_A$$
$$v_2 = i_2 R_2$$
$$= R_2 (i_A - i_B) = (5 \text{ k}\Omega)(i_A - i_B)$$
$$v_3 = i_3 R_3$$
$$= R_3 i_B = (5 \text{ k}\Omega) i_B$$

Now you can substitute the preceding voltage values into the KVL equations you found earlier:

Mesh A:
$$v_1 + v_2 = 15 \text{ V}$$
$$(10 \text{ k}\Omega)i_A + (5 \text{ k}\Omega)(i_A - i_B) = 15 \text{ V}$$

Mesh B:
$$-v_2 + v_3 = -10 \text{ V}$$
$$-(5 \text{ k}\Omega)(i_A - i_B) + (5 \text{ k}\Omega)i_B = -10 \text{ V}$$

When you rearrange the preceding equations to put them in standard form, you get

Mesh A: $(15 \text{ k}\Omega)i_A - (5 \text{ k}\Omega)i_B = 15 \text{ V}$

Mesh B: $(-5 \text{ k}\Omega)i_A + (10 \text{ k}\Omega)i_B = -10 \text{ V}$

Converting these mesh equations into matrix form results in

$$\begin{bmatrix} 15 \text{ k}\Omega & -5 \text{ k}\Omega \\ -5 \text{ k}\Omega & 10 \text{ k}\Omega \end{bmatrix} \begin{bmatrix} i_A \\ i_B \end{bmatrix} = \begin{bmatrix} 15 \text{ V} \\ -10 \text{ V} \end{bmatrix}$$

The preceding equation has the form $Ax = b$, where matrix A is the coefficients of resistors, x is a vector of unknown mesh currents, and b is a vector of independent voltage sources.

You can use your graphing calculator or matrix software to give you the mesh currents:

$$\begin{bmatrix} i_A \\ i_B \end{bmatrix} = \begin{bmatrix} 0.8 \text{ mA} \\ -0.6 \text{ mA} \end{bmatrix}$$

With these calculated mesh currents, you can find the device currents:

$$i_1 = i_A = 0.8 \text{ mA}$$
$$i_2 = i_A - i_B = 0.8 \text{ mA} - (-0.6 \text{ mA}) = 1.4 \text{ mA}$$
$$i_3 = i_B = -0.6 \text{ mA}$$

To complete the analysis, plug the device currents and resistances into the Ohm's law equations. You find the following device voltages:

$$v_1 = i_1 R_1 = (0.8 \text{ mA})(10 \text{ k}\Omega) = 8 \text{ V}$$
$$v_2 = i_2 R_2 = (1.4 \text{ mA})(5 \text{ k}\Omega) = 7 \text{ V}$$
$$v_3 = i_3 R_3 = (-0.6 \text{ mA})(5 \text{ k}\Omega) = -3 \text{ V}$$

The preceding device voltages make sense because they satisfy KVL for each mesh.

Analyzing circuits with three or more meshes

You can apply mesh-current analysis when dealing with circuits that have three or more meshes. The process is the same as for circuits with only two mesh currents. To see what I mean, consider Figure 6-5, which shows voltages and currents for each of the devices as well as the mesh currents i_A, i_B, and i_C. I chose to have all the mesh currents flow clockwise.

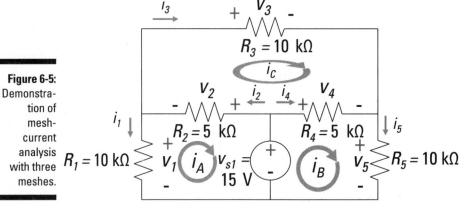

Figure 6-5: Demonstration of mesh-current analysis with three meshes.

The KVL equations for Meshes A, B, and C are

Mesh A : $v_1 + v_2 = 15$ V

Mesh B : $v_4 + v_5 = 15$ V

Mesh C : $v_2 + v_3 = v_4$

Now express the device currents in terms of mesh currents (see the earlier section "Relating Device Currents to Mesh Currents"). Then apply Ohm's law to get the element voltages in terms of the mesh currents:

$i_1 = -i_A$ → $v_1 = (10 \text{ k}\Omega)i_1 = -(10 \text{ k}\Omega)i_A$

$i_2 = i_C - i_A$ → $v_2 = (5 \text{ k}\Omega)i_2 = (5 \text{ k}\Omega)(i_C - i_A)$

$i_3 = i_C$ → $v_3 = (10 \text{ k}\Omega)i_3 = (10 \text{ k}\Omega)i_C$

$i_4 = i_B - i_C$ → $v_4 = (5 \text{ k}\Omega)i_4 = (5 \text{ k}\Omega)(i_B - i_C)$

$i_5 = i_B$ → $v_5 = (10 \text{ k}\Omega)i_5 = (10 \text{ k}\Omega)i_B$

When you substitute the preceding device voltages into the KVL equations found earlier, you wind up with

Mesh A: $v_1 + v_2 = v_{s1}$ $\rightarrow$ $-(10 \text{ k}\Omega)i_A + (5 \text{ k}\Omega)(i_C - i_A) = v_{s1}$

Mesh B: $v_4 + v_5 = v_{s1}$ $\rightarrow$ $(5 \text{ k}\Omega)(i_B - i_C) + (10 \text{ k}\Omega)i_B = v_{s1}$

Mesh C: $v_2 + v_3 = v_4$ $\rightarrow$ $(5 \text{ k}\Omega)(i_C - i_A) + (10 \text{ k}\Omega)i_C = (5 \text{ k}\Omega)(i_B - i_C)$

Rearrange the equations to put them in standard form. I've inserted some zeros as placeholder terms to help you set up the matrices in the next step:

Mesh A: $(10 \text{ k}\Omega + 5 \text{ k}\Omega)i_A + 0i_B - (5 \text{ k}\Omega)i_c = -v_{s1}$

Mesh B: $0i_A + (10 \text{ k}\Omega + 5 \text{ k}\Omega)i_B - (5 \text{ k}\Omega)i_c = v_{s1}$

Mesh C: $-(5 \text{ k}\Omega)i_A - (5 \text{ k}\Omega)i_B + (5 \text{ k}\Omega + 10 \text{ k}\Omega + 5 \text{ k}\Omega)i_c = 0$

And you can translate these standard-form equations into matrix form to get

$$\begin{bmatrix} 10 \text{ k}\Omega + 5 \text{ k}\Omega & 0 & -5 \text{ k}\Omega \\ 0 & 5 \text{ k}\Omega + 10 \text{ k}\Omega & -5 \text{ k}\Omega \\ -5 \text{ k}\Omega & -5 \text{ k}\Omega & 5 \text{ k}\Omega + 10 \text{ k}\Omega + 5 \text{ k}\Omega \end{bmatrix} \begin{bmatrix} i_A \\ i_B \\ i_C \end{bmatrix} = \begin{bmatrix} -15 \text{ V} \\ 15 \text{ V} \\ 0 \end{bmatrix}$$

Simplify the elements in the resistor matrix:

$$\begin{bmatrix} 15 \text{ k}\Omega & 0 & -5 \text{ k}\Omega \\ 0 & 15 \text{ k}\Omega & -5 \text{ k}\Omega \\ -5 \text{ k}\Omega & -5 \text{ k}\Omega & 20 \text{ k}\Omega \end{bmatrix} \begin{bmatrix} i_A \\ i_B \\ i_C \end{bmatrix} = \begin{bmatrix} -15 \text{ V} \\ 15 \text{ V} \\ 0 \end{bmatrix}$$

Notice that in the resistor matrix, the main-diagonal values are all positive, the off-diagonal values are all negative or zero, and the off-diagonal values are symmetric. For a circuit with an independent source, that symmetry with respect to the main diagonal is a good sign that you've set up the problem correctly.

You can use your graphing calculator or matrix software to find the mesh currents:

$$\begin{bmatrix} i_A \\ i_B \\ i_C \end{bmatrix} = \begin{bmatrix} -1.0 \text{ mA} \\ 1.0 \text{ mA} \\ 0.0 \text{ mA} \end{bmatrix}$$

The current $i_C = 0$ makes sense due to the circuit symmetry. With these calculated values for mesh currents, you find the following device currents:

$$i_1 = -i_A = -(-1.0 \text{ mA}) = 1.0 \text{ mA}$$
$$i_2 = i_C - i_A = 0 - (-1.0 \text{ mA}) = 1.0 \text{ mA}$$
$$i_3 = i_C = 0.0 \text{ mA}$$
$$i_4 = i_B - i_C = 1.0 \text{ mA} - 0 = 1.0 \text{ mA}$$
$$i_5 = i_B = 1.0 \text{ mA}$$

To complete the analysis, calculate the device voltages using Ohm's law, relating the device currents and voltages:

$$v_1 = R_1 i_1 = (10 \text{ k}\Omega) \cdot (1 \text{ mA}) \quad \rightarrow \quad v_1 = 10 \text{ V}$$
$$v_2 = R_2 i_2 = (5 \text{ k}\Omega) \cdot (1 \text{ mA}) \quad \rightarrow \quad v_2 = 5 \text{ V}$$
$$v_3 = R_3 i_3 = (10 \text{ k}\Omega) \cdot (0 \text{ mA}) \quad \rightarrow \quad v_3 = 0 \text{ V}$$
$$v_4 = R_4 i_4 = (5 \text{ k}\Omega) \cdot (1 \text{ mA}) \quad \rightarrow \quad v_4 = 5 \text{ V}$$
$$v_5 = R_5 i_5 = (10 \text{ k}\Omega) \cdot (1 \text{ mA}) \quad \rightarrow \quad v_5 = 10 \text{ V}$$

The preceding results make sense because they satisfy the KVL equations for the three meshes.

Chapter 7

Solving One Problem at a Time Using Superposition

In This Chapter

▶ Describing the superposition method

▶ Taking care of sources one at a time

▶ Getting a handle on analyzing a circuit with two independent sources

▶ Solving a circuit with three independent sources

*T*he method of circuit analysis known as *superposition* can be your best friend when you're faced with circuits that have lots of voltage and current sources. Superposition allows you to break down complex linear circuits composed of multiple independent sources into simpler circuits that have just one independent source. The total output, then, is the algebraic sum of individual outputs from each independent source. In this chapter, I show you just how superposition works. I also walk you through focusing on a single independent source when you have multiple sources in a circuit and using super-position to analyze circuits with two or three independent sources.

Discovering How Superposition Works

Superposition states that the output (or response) in any device of a linear circuit having two or more independent sources is the sum of the individual outputs resulting from each input source with all other sources turned off.

To use the superposition method, you need to understand the additive property of linearity. Linearity allows you to predict circuit behavior when applying an independent input source such as a battery. The circuit outputs (either a current or voltage for a particular device) are simply linear combinations of the independent input sources. Two properties are needed to describe linearity: proportionality and addivity.

Making sense of proportionality

The following equation describes proportionality mathematically, with x as the input transformed into some output y by a mathematical operation described as $T(x)$ scaled by a constant K:

$$y = KT(x) \quad (\text{Proportionality Property})$$

A transformation T means a mathematical function like multiplication, division, differentiation, or integration. In terms of circuits, this equation means that if you have a new input voltage that is doubled in amplitude from the original, the new circuit output will also be doubled. For example, suppose output y_1 is related to input x_1 by the following transformation

$$y_1 = T(x_1)$$

When you apply a new input $x = 2x_1$, which is twice as big as x_1, the new output y is also twice as big as the original y_1:

$$\text{New input, } x = 2x_1: \qquad y = T(x)$$
$$y = T(2x_1) = 2\underbrace{T(x_1)}_{=y_1}$$

$$\text{New output:} \qquad y = 2y_1$$

Figure 7-1 diagrams the proportionality concept. The top diagram illustrates how the output results from a transformed input. The bottom diagram shows that scaling your input by a constant K results in an output scaled by the same amount as the original output.

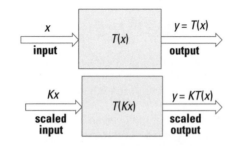

Figure 7-1: Diagram of the proportionality property.

Multiplying the input by an amount K multiplies your output by K

Voltage and current division techniques (see Chapter 4) rely on the concept of proportionality. Voltage division involves a series circuit with a voltage source. In the voltage divider method, you proportionally multiply the input voltage source v_s according to the value of a resistor and divide by the total resistance found in the series circuit to get the output voltage v_o of the resistor. Current division involves a parallel circuit with a current source. In the current division method, you proportionally multiply the input current source i_s according to the value of a conductance (or resistance) and divide by the total conductance found for the parallel circuit to get the output current i_o of the resistor.

Figure 7-2 illustrates the voltage divider technique as a proportionality concept. The top diagram illustrates a series circuit, and the bottom diagram corresponds to its block diagram viewed from an input-output system perspective. The input voltage is v_s, and the output voltage v_o is the voltage across resistor R_2 in series with resistor R_1. The output voltage is given as

$$v_o = \underbrace{\left(\frac{R_2}{R_1 + R_2} \right)}_{=K} v_s$$

$$v_o = Kv_s$$

The preceding equation shows that the output voltage v_o is proportional to the voltage source v_s scaled by the constant K.

Figure 7-2:
A voltage divider circuit and its associated system block diagram.

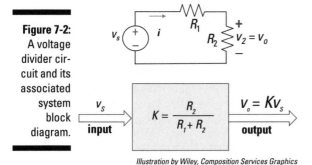

Illustration by Wiley, Composition Services Graphics

Figure 7-3 illustrates the current divider technique as a proportionality concept. The top diagram shows a parallel circuit, and the bottom diagram corresponds to its block diagram from an input-output system perspective. The input current source is i_s, and the output current i_o is the current through resistor R_2. Using current division techniques, you wind up with the following output current:

$$i_o = \underbrace{\left(\frac{R_1}{R_1 + R_2} \right)}_{=K} i_s$$

$$i_o = Ki_s$$

The preceding equation shows that the output current i_o is proportional to the current source i_s scaled by the constant K.

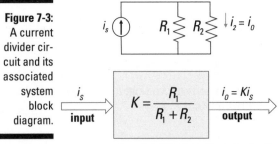

Illustration by Wiley, Composition Services Graphics

Applying superposition in circuits

You can apply the superposition (additive) property to circuits. For circuits, this property states that you can express the output current or voltage of a linear circuit having multiple inputs as a linear combination of these inputs.

Here are the steps for superposition in plain English:

1. **Find the individual output of the circuit resulting from a single source acting alone by "turning off" all other independent sources.**

   To turn off an independent source, you replace it with something that has equivalent current and voltage (i-v) characteristics:

   • **Remove an ideal voltage source by replacing it with a short circuit.** You can make this replacement because the voltage is constant in both cases, so the resistance is zero:

   $$R = \frac{\Delta v}{\Delta i} = \frac{0}{\Delta i} = 0$$

- **Replace an ideal current source with an open circuit.** You can make this replacement because the current is constant in both cases, so the resistance is infinite:

$$R = \frac{\Delta v}{\Delta i} = \frac{\Delta v}{0} = \infty$$

2. **Repeat Step 1 for each independent source to find each source's output contribution when the other sources are turned off.**

3. **Algebraically add up all the individual outputs from the sources to get the total output.**

To illustrate superposition, suppose you have an input x consisting of inputs x_1, x_2, and x_3 added together:

$$x = x_1 + x_2 + x_3$$

The superposition property states that for some transformation T operating on an input x, you obtain an output y as the sum of individual outputs due to each input, x_1, x_2, and x_3. You can mathematically describe the superposition concept as follows:

$$y = T(x_1 + x_2 + x_3)$$
$$y = \underbrace{T(x_1)}_{=y_1} + \underbrace{T(x_2)}_{=y_2} + \underbrace{T(x_3)}_{=y_3}$$

$$y = y_1 + y_2 + y_3 \quad (\text{Superposition Property})$$

In the preceding equation, the total output y is a result of three outputs: Output y_1 is due to input x_1, output y_2 is due to input x_2, and output y_3 is due to input x_3.

If you're a visual learner, take a look at Figure 7-4, which illustrates the superposition concept. In this figure, you have three inputs resulting in three outputs. Each input is transformed by the transformation T to produce an individual output. Adding up the individual outputs, you wind up with the total output y.

Figure 7-4: Diagram of the superposition concept.

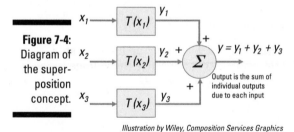

Illustration by Wiley, Composition Services Graphics

Adding the contributions of each independent source

To understand the property of superposition, consider analyzing a circuit with two independent sources: one current source and one voltage source. This example shows that the output consists of a linear combination of the current source and the voltage source.

You can see a typical circuit with two sources in Figure 7-5. Apply Kirchhoff's current law (KCL), which says that the sum of the incoming currents is equal to the sum of the outgoing currents at any node. Applying KCL at Node A produces the following result:

$$\text{in} = \text{out}: i_1 + i_s = i_2 \quad \rightarrow \quad i_1 + i_s - i_2 = 0$$

Using Ohm's law ($v = iR$), the current i_1 through R_1 is

$$i_1 = \frac{v_1}{R_1} = \frac{v_s - v_A}{R_1}$$

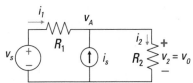

Illustration by Wiley, Composition Services Graphics

In Figure 7-5, because i_s and R_2 are connected in parallel, you have the output voltage v_o equal to the voltage v_2 (across resistor R_2). The current i_2 through R_2 using Ohm's law is

$$i_2 = \frac{v_2}{R_2} = \frac{v_o}{R_2}$$

and $v_o = v_A$. Substituting expressions for the element currents i_1 and i_2 into the KCL equation produces

$$\frac{v_s - v_o}{R_1} + i_s - \frac{v_o}{R_2} = 0$$

Algebraically move the input sources v_s and i_s to the right side of the equation, which results in

$$\left(\frac{1}{R_1}+\frac{1}{R_2}\right)v_o = \frac{v_s}{R_1}+i_s$$

Let $1/R_{eq} = 1/R_1+1/R_2$, where R_{eq} is the equivalent resistance for parallel resistors, and then solve for v_o. (In this case, the resistors are viewed as a parallel connection only when you turn off the two independent sources.) You wind up with the following output voltage v_o:

$$v_o = \underbrace{\frac{R_{eq}}{R_1}}_{=K_1}v_s + \underbrace{R_{eq}}_{=K_2}i_s$$
$$v_o = K_1 v_s + K_2 i_s$$

This equation shows output v_o as a linear combination of two input sources, v_s and i_s.

Getting Rid of the Sources of Frustration

To apply the superposition technique, you need to turn off independent sources so you can look at the output contribution from each input source. To see how to turn off independent sources, you look at the current and voltage (*i-v*) characteristics of a voltage source and a current source, as I explain in the next sections. Although *i-v* graphs are typically drawn with *i* as the vertical axis and *v* as the horizontal axis, as I show you in Chapter 2, I've reversed that pattern in the graphs in the following sections simply to show that the slope is zero for a resistor of zero.

Short circuit: Removing a voltage source

To remove an ideal voltage source, replace it with a short circuit, because both a short circuit and an ideal voltage source have zero resistance. To see why an ideal voltage source has zero resistance, look at the *i-v* characteristic of an ideal voltage source in Figure 7-6.

The ideal voltage source has a constant voltage, regardless of the current supplied by the voltage source. Because the voltage source has constant voltage, the slope of the resistance is zero. The slope is given as

$$\text{slope} = \frac{\Delta v}{\Delta i} = R$$

An ideal voltage source doesn't change in the voltage for a given change in the supplied current, so $\Delta v = 0$, which implies that the slope is zero. Mathematically,

$$\text{Slope } R: \quad R = \frac{\Delta v}{\Delta i} = \frac{0}{\Delta i} \quad \rightarrow \quad R = 0 \quad (\text{short circuit})$$

In other words, turning off the voltage source means replacing it with a short circuit. Using a short circuit ensures that no voltage is present from the voltage source.

Figure 7-6: Turning off an ideal voltage source with a short circuit (zero resistance).

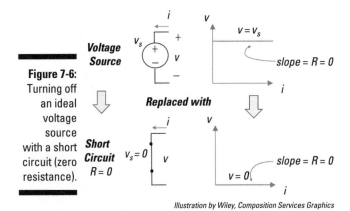

Illustration by Wiley, Composition Services Graphics

Open circuit: Taking out a current source

To remove an ideal current source, replace it with an open circuit, because both devices have infinite resistance. To see why a current source has infinite resistance, look at its current and voltage (*i-v*) characteristic.

The ideal current source provides constant current, regardless of the voltage across the current source; that is, $\Delta i = 0$. With constant current, the slope of the resistance is infinite. With R as the slope of the line in Figure 7-7, the slope is mathematically

$$\text{Slope } R: \quad R = \frac{\Delta v}{\Delta i} = \frac{\Delta v}{0} = \infty \quad \rightarrow \quad R = \infty \quad (\text{open circuit})$$

In other words, turning off the current source means removing the current source from the circuit or, equivalently, replacing the current source with an open circuit. Using an open circuit ensures that no current flows.

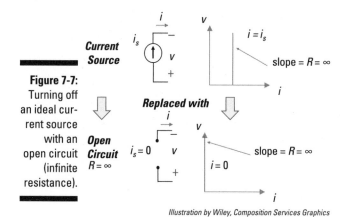

Analyzing Circuits with Two Independent Sources

The simplest way to understand superposition is to tackle circuits that have just two independent sources. That's why the following sections focus on such circuits. One circuit has two independent voltage sources, another circuit has two independent current sources, and the last has both voltage and current sources.

Knowing what to do when the sources are two voltage sources

With the help of superposition, you can break down the complex circuit in Figure 7-8 into two simpler circuits that have just one voltage source each. To turn off a voltage source, you replace it with a short circuit.

Circuit A contains two voltage sources, v_{s1} and v_{s2}, and you want to find the output voltage v_o across the 10-kΩ resistor. The next diagram shows the same circuit with one voltage source turned off: Circuit B contains one voltage source, with v_{s2} turned off and replaced by a short circuit. The output voltage due to v_{s1} is v_{o1}. Similarly, Circuit C is Circuit A with the other voltage source turned off. Circuit C contains one voltage source, with v_{s1} replaced by a short circuit. The output voltage due to voltage source v_{s2} is v_{o2}.

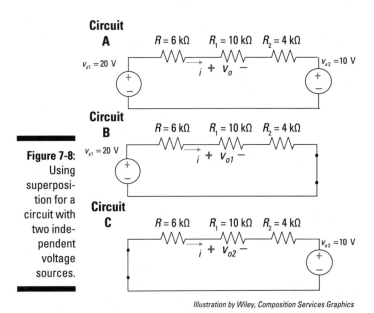

Figure 7-8: Using superposition for a circuit with two independent voltage sources.

Summing up the two outputs due to each voltage source, you wind up with the following output voltage:

$$v_o = v_{o1} + v_{o2}$$

To find the output voltages for Circuits B and C, you use voltage divider techniques. That is, you use the idea that a circuit with a voltage source connected in series with resistors divides its source voltage proportionally according to the ratio of a resistor value to the total resistance. (For the full scoop on the voltage divider technique, see Chapter 4.)

In Circuit B, you simply find the output voltage v_{o1} due to v_{s1} with a voltage divider equation:

$$v_{o1} = v_{s1}\left(\frac{R_1}{R + R_1 + R_2}\right)$$

$$v_{o1} = (20 \text{ V})\left(\frac{10 \text{ k}\Omega}{6 \text{ k}\Omega + 10 \text{ k}\Omega + 4 \text{ k}\Omega}\right) = 10 \text{ V}$$

In Circuit C, finding the output voltage v_{o2} due to v_{s2} also requires a voltage divider equation, with the polarities of v_{o2} opposite v_{s2}. Using the voltage divider method produces the output voltage v_{o2} as follows:

$$v_{o2} = -v_{s2}\left(\frac{R_1}{R+R_1+R_2}\right)$$

$$v_{o2} = (-10 \text{ V})\left(\frac{10 \text{ k}\Omega}{6 \text{ k}\Omega + 10 \text{ k}\Omega + 4 \text{ k}\Omega}\right) = -5 \text{ V}$$

Adding up the individual outputs due to each source, you wind up with the following total output for the voltage across the 10-kΩ resistor:

$$v_o = v_{o1} + v_{o2} = (10 \text{ V} - 5 \text{ V}) = 5 \text{ V}$$

Proceeding when the sources are two current sources

The plan in this section is to reduce the circuit in Figure 7-9 to two simpler circuits, each one having a single current source, and add the outputs using superposition. You consider the outputs from the current sources one at a time, turning off a current source by replacing it with an open circuit.

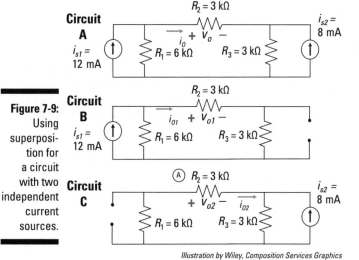

Figure 7-9: Using superposition for a circuit with two independent current sources.

Illustration by Wiley, Composition Services Graphics

Circuit A consists of two current sources, i_{s1} and i_{s2}, and you want to find the output current i_o flowing through resistor R_2. Circuit B is the same circuit with one current source turned off: Circuit B contains one current source, with i_{s2} replaced by an open circuit. The output voltage due to i_{s1} is i_{o1}. Similarly, Circuit C is Circuit A with only current source, with i_{s1} replaced by an open circuit. The output current due to current source i_{s2} is i_{o2}.

Adding up the two current outputs due to each source, you wind up with the following net output current through R_2:

$$i_o = i_{o1} + i_{o2}$$

To find the output currents for Circuits B and C, you use current divider techniques. That is, you use the idea that for a parallel circuit, the current source connected in parallel with resistors divides its supplied current proportionally according to the ratio of the value of the conductance to the total conductance. (See Chapter 4 for more on the current divider technique.)

For Circuit B, you find the output current i_{o1} due to i_{s1} using a current divider equation. Note that there are two 3-kΩ resistors connected in series in one branch of the circuit, so use their combined resistance in the equation. Given $R_{eq1} = 3$ kΩ + 3 kΩ and $R_1 = 6$ kΩ, here's output current for the first current source:

$$i_{o1} = \frac{\left(\dfrac{1}{R_{eq1}}\right)}{\left(\dfrac{1}{R_1} + \dfrac{1}{R_{eq1}}\right)} \cdot i_{s1}$$

$$i_{o1} = \left(\frac{\dfrac{1}{(3\text{ k}\Omega + 3\text{ k}\Omega)}}{\dfrac{1}{6\text{ k}\Omega} + \dfrac{1}{(3\text{ k}\Omega + 3\text{ k}\Omega)}}\right) \cdot (12\text{ mA}) = 6\text{ mA}$$

In Circuit C, the output current i_{o2} due to i_{s2} also requires a current divider equation. Note the current direction between i_{o2} and i_{s2}: i_{s2} is opposite in sign to i_{o2}. Given $R_{eq2} = 6$ kΩ + 3 kΩ and $R_3 = 3$ kΩ, the output current from the second current source is

$$i_{o2} = \frac{\left(\dfrac{1}{R_{eq2}}\right)}{\left(\dfrac{1}{R_{eq2}} + \dfrac{1}{R_3}\right)} \cdot i_{s2}$$

$$i_{o2} = \left(\frac{\dfrac{1}{(6\text{ k}\Omega + 3\text{ k}\Omega)}}{\dfrac{1}{(6\text{ k}\Omega + 3\text{ k}\Omega)} + \dfrac{1}{3\text{ k}\Omega}}\right) \cdot (-8\text{ mA}) = -2\text{ mA}$$

Adding up i_{o1} and i_{o2}, you wind up with the following total output current:

$$i_o = i_{o1} + i_{o2} = 6\text{ mA} - 2\text{ mA} = 4\text{ mA}$$

Dealing with one voltage source and one current source

You can use superposition when a circuit has a mixture of two independent sources, with one voltage source and one current source. You need to turn off the independent sources one at a time. To do so, replace the current source with an open circuit and the voltage source with a short circuit.

Circuit A of Figure 7-10 has an independent voltage source and an independent current source. How do you find the output voltage v_o as the voltage across resistor R_2?

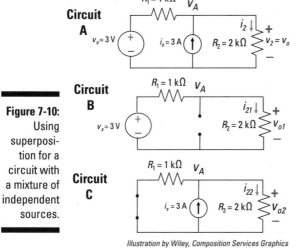

Figure 7-10:
Using
superposi-
tion for a
circuit with
a mixture of
independent
sources.

Illustration by Wiley, Composition Services Graphics

Circuit A (with its two independent sources) breaks up into two simpler circuits, B and C, which have just one source each. Circuit B has one voltage source because I replaced the current source with an open circuit. Circuit C has one current source because I replaced the voltage source with a short circuit.

For Circuit B, you can use the voltage divider technique because its resistors, R_1 and R_2, are connected in series with a voltage source. So here's the voltage v_{o1} across resistor R_2:

$$v_{o1} = \left(\frac{R_2}{R_1 + R_2} \right) v_s$$

$$v_{o1} = \left(\frac{2 \text{ k}\Omega}{1 \text{ k}\Omega + 2 \text{ k}\Omega} \right)(3 \text{ V}) = 2 \text{ V}$$

For Circuit C, you can use a current divider technique because the resistors are connected in parallel with a current source. The current source provides the following current i_{22} flowing through resistor R_2:

$$i_{22} = \left(\frac{R_1}{R_1 + R_2} \right) i_s$$

$$i_{22} = \left(\frac{1 \text{ k}\Omega}{1 \text{ k}\Omega + 2 \text{ k}\Omega} \right) (3 \text{ A}) = 1 \text{ mA}$$

You can use Ohm's law to find the voltage output v_{o2} across resistor R_2:

$$v_{o2} = i_{22} R_2$$

$$v_{o2} = (1 \text{ mA})(2 \text{ k}\Omega) = 2 \text{ V}$$

Now find the total output voltage across R_2 for the two independent sources in Circuit C by adding v_{o1} (due to the source voltage v_s) and v_{o2} (due to the source current i_s). You wind up with the following output voltage:

$$v_o = v_{o1} + v_{o2}$$

$$v_o = 2 \text{ V} + 2 \text{ V} = 4 \text{ V}$$

Solving a Circuit with Three Independent Sources

You can use superposition when faced with a circuit that has three (or more) independent sources. With three independent sources, you find the output voltage of three simplified circuits, where each circuit has one source working and the others turned off. Then add the outputs due to the three power sources.

Circuit A in Figure 7-11 has two voltage sources and one current source. Suppose you want to find the output voltage across the current source i_s.

To help you follow the analysis, I identified the voltage v_{AB} by labeling Terminals A and B. This voltage is equal to the output voltage v_o across the current source. The voltage across the current source is equivalent to the voltage across resistor R_3 connected in series with voltage source v_{s2}.

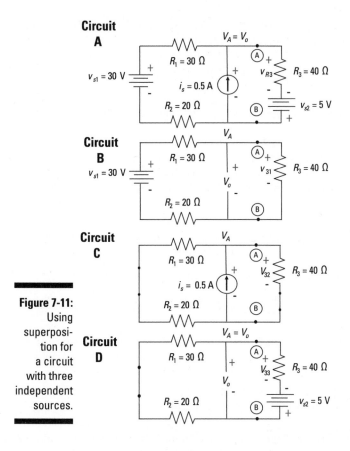

Circuit A

$V_A = V_o$
$R_1 = 30\ \Omega$
$V_{s1} = 30\ V$
$i_s = 0.5\ A$
V_{R3}
$R_3 = 40\ \Omega$
$R_2 = 20\ \Omega$
$V_{s2} = 5\ V$

Circuit B

V_A
$R_1 = 30\ \Omega$
$V_{s1} = 30\ V$
V_{31}
$R_3 = 40\ \Omega$
V_o
$R_2 = 20\ \Omega$

Circuit C

V_A
$R_1 = 30\ \Omega$
$i_s = 0.5\ A$
V_{32}
$R_3 = 40\ \Omega$
$R_2 = 20\ \Omega$

Circuit D

$V_A = V_o$
$R_1 = 30\ \Omega$
V_{33}
$R_3 = 40\ \Omega$
V_o
$R_2 = 20\ \Omega$
$V_{s2} = 5\ V$

Figure 7-11:
Using superposition for a circuit with three independent sources.

In Circuit A, the voltage across the current source i_s is connected in parallel with the series combination of R_3 and v_{s2}. You can find the voltage across R_3 and v_{s2}, which is equal to the output voltage v_o.

Applying Kirchhoff's voltage law (KVL) to describe this situation, you wind up with

$$v_o = v_{R3} + v_{s2} = v_{R3} - 5\ V$$

Essentially, finding v_o involves finding the voltage across resistor R_3. When you know this voltage, you can easily calculate the output voltage, v_o, with the preceding equation.

You can break down Circuit A, with three independent sources, into simpler Circuits B, C, and D, each having a single independent source with the other sources removed or turned off. To analyze the simpler circuits with one source, you apply voltage and current divider techniques, which I introduce in Chapter 4.

You need to first find the voltage across R_3 due to each independent source. Here's how it works:

- ✔ **Source 1: Circuit B, first voltage source:** You calculate the voltage across R_3 due to v_{s1} by first removing the voltage source v_{s2} and replacing it with a short. You also remove the current source i_s by replacing it with an open circuit.

  After removing two independent sources, you have Circuit B, a series circuit driven by a single voltage source, v_{s1}. Consequently, the voltage divider technique applies, yielding a voltage v_{31} across resistor R_3 due to v_{s1}:

  $$v_{31} = (30 \text{ V})\left(\frac{40 \ \Omega}{30 \ \Omega + 20 \ \Omega + 40 \ \Omega}\right) = 13.33 \text{ V}$$

- ✔ **Source 2: Circuit C, current source:** You calculate the voltage across R_3 due to i_s by first removing the voltage sources v_{s1} and v_{s2} and replacing them with shorts.

  After removing two independent voltage sources, you have Circuit C, a parallel circuit driven by a single current source i_s. As a result, the current divider technique applies. This produces a current i_{32} through resistor R_3, resulting from current source i_s. Also not that the voltage polarity of V_{s2} is opposite that of v_{33}. Using the current divider for Circuit C yields the following:

  $$i_{32} = (0.50 \text{ A})\left(\frac{50 \ \Omega}{50 \ \Omega + 40 \ \Omega}\right) = 0.2778 \text{ A}$$

  Next, use Ohm's law to find the voltage across R_3 due to current source i_s:

  $$v_{32} = (0.278 \text{ A})(40 \ \Omega) = 11.11 \text{ V}$$

- ✔ **Source 3: Circuit D, second voltage source:** You calculate the voltage across R_3 due to v_{s2} by first removing the voltage source v_{s1} and replacing it with a short circuit. Also remove the current source i_s by replacing it with an open circuit.

After removing two independent sources, you have Circuit D, a series circuit driven by a single voltage source, v_{s2}. Because this is a series circuit, the voltage divider technique applies, producing a voltage v_{33} across resistor R_3 due to v_{s2}. Also note that the voltage polarity of v_{s2} is opposite that of v_{33}. Using the voltage divider technique produces the following output:

$$v_{33} = (5 \text{ V})\left(\frac{40 \ \Omega}{30 \ \Omega + 20 \ \Omega + 40 \ \Omega}\right) = 2.222 \text{ V}$$

To find v_{R3}, add up the voltages across resistor R_3 due to each independent source:

$v_{R3} = v_{31} + v_{32} + v_{33}$

$v_{R3} = 13.33 \text{ V} + 11.11 \text{ V} + 2.222 \text{ V}$

$v_{R3} = 26.67 \text{ V}$

Here's the total output voltage $(v_o + v_{AB})$ across the current source (or voltage v_{AB} across Terminals A and B):

$V_o = V_{AB} = V_{R3} - V_{s2} = 26.67 \text{ V} - 5 \text{ V} = 21.67 \text{ V}$

Chapter 8

Applying Thévenin's and Norton's Theorems

In This Chapter

▶ Simplifying source circuits with Thévenin's and Norton's theorems

▶ Using the Thévenin and Norton approach with superposition

▶ Delivering maximum power transfer

*T*hévenin and Norton equivalent circuits are valuable tools when you're connecting and analyzing two different parts of a circuit. In this chapter, one part of the circuit, called the *source circuit,* delivers signals and interacts with another part, dubbed a *load circuit.* The interaction between the source and load circuits offers a major challenge when analyzing circuits.

Fortunately, Thévenin's theorem and Norton's theorem simplify the analysis. Each theorem allows you to replace a complicated array of independent sources and resistors, turning the source circuit into a single independent source connected with a single resistor. As a result, you don't have to reanalyze the entire circuit when you want to try different loads — you can just use the same source circuit.

This chapter reveals just how Thévenin's and Norton's theorems work. It also shows you how to use the superposition technique (which I present in Chapter 7) to find equivalent Thévenin and Norton circuits when a circuit has multiple sources. Last but not least, I explain how to apply the Thévenin or Norton equivalent to show how to deliver maximum power to a load circuit.

Showing What You Can Do with Thévenin's and Norton's Theorems

Trying out and replacing devices in a circuit can be dull and dreary work. But you can minimize the toil by replacing part of the circuit with a simpler but equivalent circuit using Thévenin's or Norton's theorem.

These theorems come in handy when you're faced with circuits like the one in Figure 8-1. Note the following parts of the circuit:

- ✔ The circuit to the left of Terminals A and B is the *source circuit.* It's a linear circuit with an array of voltage sources and resistors.

- ✔ The source circuit delivers a signal to the *load circuits,* which are to the right of Terminals A and B.

- ✔ The terminals at A and B make up the *interface* between the source circuit and the load circuits.

To find the voltage across each device in the load circuit, you'd usually have to connect Devices 1, 2, and 3 one at a time to get three different answers for three different loads. Talk about tedious!

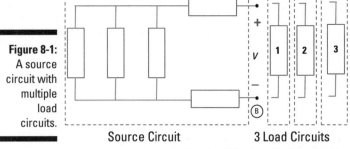

Figure 8-1:
A source circuit with multiple load circuits.

Source Circuit   3 Load Circuits

Illustration by Wiley, Composition Services Graphics

Here's where Thévenin's theorem comes to the rescue. *Thévenin's theorem* lets you replace the source circuit — a linear array of devices having multiple independent sources and resistors — with a single voltage source connected in series with a single resistor. Figure 8-2 replaces the source circuit from Figure 8-1 with a simplified circuit, which has a Thévenin voltage source, v_T, connected in series with a Thévenin resistor, R_T. The Thévenin equivalent is useful when the devices in the load circuits are connected in series.

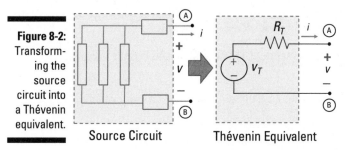

Figure 8-2:
Transforming the source circuit into a Thévenin equivalent.

Source Circuit   Thévenin Equivalent

Illustration by Wiley, Composition Services Graphics

Norton's theorem likewise allows you to simplify the source circuit. Specifically, *Norton's theorem* says you can replace a linear array of devices having multiple independent sources and resistors with a single current source in parallel with a single resistor. Check out Figure 8-3 for an example. It shows the linear source circuit being replaced with a simplified circuit that has one current source, i_N, and one resistor, R_N, connected in parallel. The Norton equivalent is useful when you want to try loads that have devices connected in parallel.

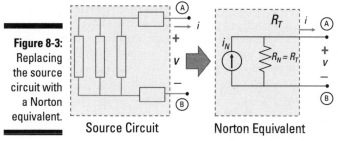

Figure 8-3:
Replacing the source circuit with a Norton equivalent.

Source Circuit   Norton Equivalent

Illustration by Wiley, Composition Services Graphics

Finding the Norton and Thévenin Equivalents for Complex Source Circuits

Norton's theorem and Thévenin's theorem say essentially the same thing. The Norton equivalent is the Thévenin equivalent with a source transformation (I cover source transformation in Chapter 4).

REMEMBER

To find the Thévenin or Norton equivalent of a linear source circuit, calculate the following variables at interface Terminals A and B:

- ✔ **Thévenin voltage source, v_T:** This equals open-circuit voltage:

$$v_T = v_{oc}$$

- ✔ **Norton current source, i_N:** This equals short-circuit current:

$$i_N = i_{sc}$$

- ✔ **Thévenin resistance, R_T, or Norton resistance, R_N:** This resistance equals open-circuit voltage divided by short-circuit current:

$$R_T = R_N = \frac{v_{OC}}{i_{SC}}$$

So where do these equivalents come from? Resistor loads can have a wide variety of resistor values, ranging from a short circuit having zero resistance to an open circuit having infinite resistance. These extreme ends of the resistance spectrum are convenient when you're analyzing circuits because you can easily find the Thévenin voltage v_T by having an open-circuit load, and you can get the Norton current i_N by having a short-circuit load.

Figure 8-4 shows a source circuit and its Thévenin equivalent. The top diagram shows the open circuit load you use to find the open-circuit voltage, v_{oc}, across Terminals A and B. The bottom diagram shows the short-circuit load you use to find the short-circuit current, i_{sc}, through Terminals A and B. With the open-circuit voltage, v_{oc}, and the short-circuit current, i_{sc}, you can find the Thévenin or Norton resistance ($R_T = R_N$).

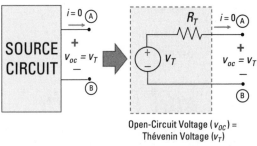

Open-Circuit Voltage (v_{OC}) = Thévenin Voltage (v_T)

Figure 8-4:
Finding the Thévenin equivalent using open-circuit loads (top) and short-circuit loads (bottom).

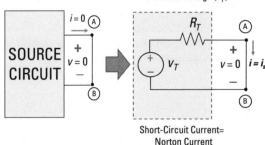

Short-Circuit Current= Norton Current

Illustration by Wiley, Composition Services Graphics

In the following sections, I show you how to apply Thévenin's and Norton's theorems, and I show you how to use source transformation to go from one equivalent to another. I also offer an alternate way of finding R_T or R_N: finding the total resistance between Terminals A and B by removing all the independent sources of the source circuit.

Applying Thévenin's theorem

To simplify your analysis when interfacing between source and load circuits, the Thévenin method replaces a complex source circuit with a single voltage source in series with a single resistor. To obtain the Thévenin equivalent, you need to calculate the open-circuit voltage v_{oc} and the short-circuit current i_{sc}.

Finding the Thévenin equivalent of a circuit with a single independent voltage source

Circuit A in Figure 8-5 is a source circuit with an independent voltage source connected to a load circuit. Circuit B shows the same circuit, except I've replaced the load circuit with an open-circuit load. You use the open-circuit load to get the Thévenin voltage, v_T, across Terminals A and B. The Thévenin voltage equals the open-circuit voltage, v_{oc}.

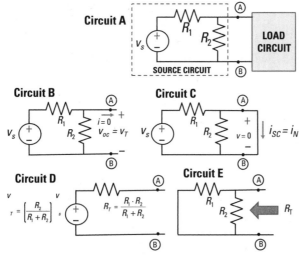

Figure 8-5: Using the Thévenin equivalent.

Illustration by Wiley, Composition Services Graphics

The voltage is driven by a voltage source for this series circuit, so use the voltage divider technique (from Chapter 4) to get v_{oc}:

$$v_{oc} = \left(\frac{R_2}{R_1 + R_2} \right) v_s \qquad \text{(open-circuit voltage)}$$

$$v_{oc} = v_T$$

Solving for v_{oc} gives you the Thévenin voltage, v_T.

Circuit C shows the same source circuit as a short-circuit load. You use the short-circuit load to get the Norton current, i_N, through Terminals A and B. And you find the Norton current by finding the short-circuit current, i_{sc}.

In Circuit C, the short circuit is in parallel with resistor R_2. This means that all the current coming out from resistor R_1 will flow through the short because the short has zero resistance. In other words, the short bypasses R_2. You can find the current through Terminals A and B using Ohm's law, producing the short-circuit current:

$$i_{sc} = \frac{v_s}{R_1}$$

$$i_{sc} = i_N$$

This short-circuit current, i_{sc}, gives you the Norton current, i_N.

Finally, to get the Thévenin resistance, R_T, you divide the open-circuit voltage by the short-circuit current. You then wind up with the following expression for R_T:

$$R_T = \frac{v_{oc}}{i_{sc}}$$

$$R_T = \frac{\left(\frac{R_2}{R_1 + R_2} \right) v_s}{\frac{v_s}{R_1}}$$

Simplify that equation to get the Thévenin resistance:

$$R_T = \frac{R_1 \cdot R_2}{R_1 + R_2}$$

Circuit D shows the Thévenin equivalent for the source circuit in Circuit A.

The preceding equation looks like the total resistance for the parallel connection between resistors R_1 and R_2 when you short (or remove) the voltage source and look back from Terminals A and B.

When looking to the left from the Terminals A and B, you can find the Thévenin resistance R_T by removing all independent sources by shorting voltage sources and replacing current sources with open circuits. After getting rid of the independent sources, you can find the total resistance between Terminals A and B, shown in Circuit E of Figure 8-5. (Note that this tactic only works when there are no dependent sources.)

Applying Norton's theorem

To see how to use the Norton approach for circuits with multiple sources, consider Circuit A in Figure 8-6. Because it doesn't matter whether you find the short-circuit current or the open-circuit voltage first, you can begin by determining the open-circuit voltage. Putting an open load at Terminals A and B results in Circuit B. The following analysis shows you how to obtain i_{s1} and R_N in Circuit B.

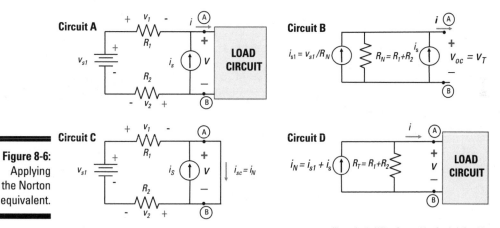

Figure 8-6:
Applying
the Norton
equivalent.

Illustration by Wiley, Composition Services Graphics

Applying Kirchhoff's voltage law (KVL) in Circuit A lets you determine the open-circuit voltage, v_{oc}. KVL says that the sum of the voltage rises and drops around the loop is zero. Assuming an open circuit load for Circuit A, you get the following KVL equation (where the load is an open circuit, $v = v_{oc}$):

$$-v_{s1} + v_1 + v_2 + v_{oc} = 0$$

Algebraically solve for v_{oc} to get the open-circuit voltage:

$$v_{oc} = v_{s1} - v_1 - v_2$$

The current supplied by the voltage source v_s goes through resistors R_1 and R_2 because the current going through an open circuit load is zero. In Circuit B, you can view the current source i_s as a device having an infinite resistance (that is, as an open circuit). However, all the current provided by the current source i_s will go through R_1 and R_2, and none of the current from i_s will go through the open-circuit load. Applying Ohm's law ($v = iR$), you have the following voltages across resistors R_1 and R_2:

$$v_1 = -i_s R_1$$
$$v_2 = -i_s R_2$$

The minus sign appears in these equations because the current from i_s flows opposite in direction to the assigned voltage polarities across the resistors.

Substitute v_1 and v_2 into the expression for v_{oc}, and you wind up with the following open-circuit voltage:

$$v_{oc} = v_{s1} + i_s (R_1 + R_2)$$

The open-circuit voltage is equal to the Thévenin equivalent voltage, $v_{oc} = v_T$.

Next, find the short-circuit current in Circuit C of Figure 8-6. The current i_{s1} supplied by the voltage source will flow only through resistors R_1 and R_2, not through the current source i_s, which has infinite resistance. Because of the short circuit, the resistors R_1 and R_2 are connected in series, resulting in an equivalent resistance of $R_1 + R_2$. Applying Ohm's law to this series combination gives you the following expression for i_{s1} provided by the voltage source v_{s1}:

$$i_{s1} = \frac{v_{s1}}{R_1 + R_2}$$

Kirchhoff's current law (KCL) says that the sum of the incoming currents is equal to the sum of the outgoing currents at a node. Applying KCL at Node A, you get

$$i_{s1} + i_s = i_{sc}$$

Substituting the expression for i_{s1} into the preceding KCL equation gives you the short-circuit current, i_{sc}:

$$i_{sc} = \frac{v_{s1}}{R_1 + R_2} + i_s = i_N$$

The Norton current i_N is equal to the short-circuit current: $i_N = i_{sc}$.

Finally, divide the open-circuit voltage by the short-circuit current to get the Norton resistance, R_N:

$$R_N = R_T = \frac{v_{oc}}{i_{sc}}$$

Plugging in the expressions for v_{oc} and i_{sc} gives you the Norton resistance:

$$R_N = \frac{v_{s1} + i_s(R_1 + R_2)}{\dfrac{v_{s1}}{R_1 + R_2} + i_s}$$

Adding the terms in the denominator requires adding fractions, so rewrite the terms so they have a common denominator. Algebraically, the equation simplifies as follows:

$$R_N = \frac{v_{s1} + i_s(R_1 + R_2)}{\left(\dfrac{v_{s1} + i_s(R_1 + R_2)}{R_1 + R_2}\right)}$$

$$R_N = \left(v_{s1} + i_s(R_1 + R_2)\right)\left(\frac{R_1 + R_2}{v_{s1} + i_s(R_1 + R_2)}\right)$$

$$R_N = R_1 + R_2$$

When you look left from the right of Terminals A and B, the Norton resistance is equal to the total resistance while removing all the independent sources. You see the Norton equivalent in Circuit D of Figure 8-6, where $R_T = R_N$.

Using source transformation to find Thévenin or Norton

In this section, I show you how to apply Thévenin's and Norton's theorems to analyze complex circuits using source transformation.

You commonly use the Thévenin equivalent when you want circuit devices to be connected in series and the Norton equivalent when you want devices to be connected in parallel with the load. You can then use the voltage divider technique for a series circuit to obtain the load voltage, or you can use the current divider technique for a parallel circuit to obtain the load current.

A shortcut: Finding Thévenin or Norton equivalents with source transformation

To transform a circuit using the Thévenin or Norton approach, you need to know both the Thévenin voltage (open-circuit voltage) and the Norton current (short-circuit current). But you don't need to find the Thévenin and Norton equivalent circuits separately. After you figure out the Thévenin equivalent for a circuit, you can find the Norton equivalent using source transformation. Or if you figure out the Norton equivalent first, source transformation lets you find the Thévenin equivalent. (I cover source transformations in Chapter 4.)

For example, if you already have the Thévenin equivalent circuit, then obtaining the Norton equivalent is a piece of cake. You perform the source transformation to convert the Thévenin voltage source connected in series with the Thévenin resistance into a current source connected in parallel with the Thévenin resistance. The result is the Norton equivalent: The current source is the Norton current source, and the Thévenin resistance is the Norton resistance.

Finding the Thévenin equivalent of a circuit with multiple independent sources

You can use the Thévenin approach for circuits that have multiple independent sources. In some cases, you can use source transformation techniques to find the Thévenin resistor R_T without actually computing v_{oc} and i_{sc}.

For example, consider Circuit A in Figure 8-7. In this circuit, the voltage source v_s and resistors R_1 and R_2 are connected in series. When you remove independent sources v_{S1} and i_s in Circuit A, this series combination of resistors produces the following total Thévenin resistance:

$$R_T = R_1 + R_2$$

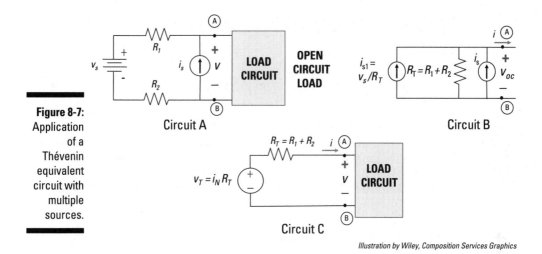

Illustration by Wiley, Composition Services Graphics

Figure 8-7: Application of a Thévenin equivalent circuit with multiple sources.

You can then use source transformation to convert the Thévenin voltage source, which is connected in series to Thévenin resistance R_T, into a current source that's connected in parallel with R_T. Here's your current source:

$$i_{s1} = \frac{v_s}{R_T}$$

Circuit B shows the transformed circuit with two independent current sources.

Because independent current sources are in parallel and point in the same direction, you can add up the two source currents, which produces the equivalent Norton current, i_N:

$$i_N = i_s + i_{s1}$$
$$i_N = i_s + \frac{v_s}{R_T}$$

Circuit B shows the combination of the two current sources. When you combine the two current sources into one single current source connected in parallel with one resistor, you have the Norton equivalent.

You can convert the current source i_N in parallel with R_T to a voltage source in series with R_T using the following source transformation equation:

$$v_T = i_N R_T$$

Circuit C is the Thévenin equivalent consisting of one voltage source connected in series with a single equivalent resistor, R_T.

Finding Thévenin or Norton with superposition

When a complex circuit has multiple sources, you can use superposition to obtain either the Thévenin or Norton equivalent. As I explain in Chapter 7, superposition involves determining the contribution of each independent source while turning off the other sources. After determining the contribution of each source, you add up the contributions of all the sources.

This section shows you how to use superposition to find the Thévenin equivalent, but the process for finding the Norton equivalent is essentially the same. You simply find the Thévenin equivalent — consisting of one voltage source connected in series with a resistor — and get the Norton equivalent through a source transformation.

To see how superposition can help you obtain the Thévenin equivalent, consider Circuit A of Figure 8-8. To find the open-circuit voltage due to only the voltage source v_s, you turn off the current source i_s by removing it from Circuit A. Circuit B is the resulting circuit.

Because of the open-circuit load in Circuit B, no current will flow through resistors R_1 and R_2. And because there's no current flow, the voltage drop across each of these two resistors is equal to zero, according to Ohm's law ($v = iR$). The open-circuit voltage due to v_s, denoted as v_{oc1}, is therefore equal to v_s. Mathematically, you can write

$$v_{oc1} = v_s$$

To find the open-circuit voltage contribution due to the current source i_s, you turn off the voltage source v_s by replacing it with a short circuit, which has zero resistance. You see the resulting circuit in Circuit C.

Because no current flows through the open-circuit load, i_s flows through resistors R_1 and R_2. The open-circuit voltage across Terminals A and B, denoted as v_{oc2}, is equal to the voltage drop across the two resistors. Using Ohm's law, you have the following expression for v_{oc2}:

$$v_{oc2} = i_s \left(R_1 + R_2 \right)$$

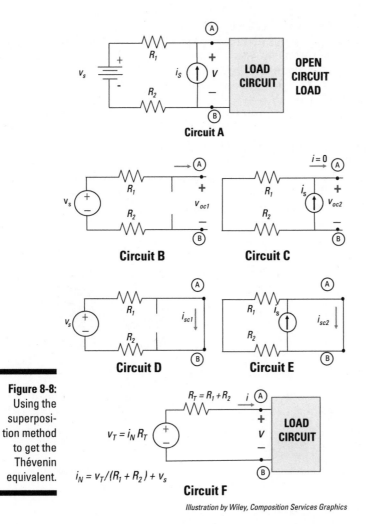

Figure 8-8:
Using the superposition method to get the Thévenin equivalent.

Circuit A

Circuit B   Circuit C

Circuit D   Circuit E

$R_T = R_1 + R_2$

$v_T = i_N R_T$

$i_N = v_T/(R_1 + R_2) + v_s$

Circuit F

Illustration by Wiley, Composition Services Graphics

Adding up v_{oc1} and v_{oc2} gives you the total open-circuit output contribution due to v_s and i_s:

$$v_{oc} = v_{oc1} + v_{oc2}$$
$$= v_s + i_s(R_1 + R_2)$$

The open-circuit voltage equals the Thévenin voltage source ($v_{oc} = v_T$).

To find the short-circuit current due to only voltage source v_s, you turn off the current source i_s by removing it from Circuit A. The result is Circuit D.

Current flows through resistors R_1 and R_2 because the short connects the resistors in series. The series combination gives you an equivalent resistance of $R_1 + R_2$. Applying Ohm's law, you get the following expression for i_{sc1}, which is the short-circuit current due to v_s:

$$i_{sc1} = \frac{v_s}{R_1 + R_2}$$

To find the short-circuit current contribution due to only current source i_s, you turn off the voltage source v_s by replacing it with a short circuit, which has zero resistance. The result is Circuit E. Because of the short circuit, all the current provided by i_s flows through Terminals A and B. In other words, i_{sc2}, which is the short-circuit current resulting from i_s, equals the source current i_s:

$$i_{sc2} = i_s$$

Adding up i_{sc1} and i_{sc2} gives you the total contribution due to v_s and i_s:

$$i_{sc} = i_{sc1} + i_{sc2}$$
$$= \frac{v_s}{R_1 + R_2} + i_s$$

This short-circuit current equals the Norton current source ($i_N = i_{sc}$).

Divide the expression for v_{oc} by the expression for i_{sc}, and you get the Thévenin resistance:

$$R_T = \frac{v_{oc}}{i_{sc}} = \frac{v_s + i_s(R_1 + R_2)}{\left(\dfrac{v_s}{R_1 + R_2} + i_s\right)}$$

This equation simplifies as follows:

$$R_T = \frac{v_s + i_s(R_1 + R_2)}{\left(\dfrac{v_s + i_s(R_1 + R_2)}{R_1 + R_2}\right)}$$

$$R_T = (v_s + i_s(R_1 + R_2))\left(\frac{R_1 + R_2}{v_s + i_s(R_1 + R_2)}\right)$$

$$R_T = R_1 + R_2$$

The Thévenin resistance is equal to the total resistance of the series combination when you're looking left from the right of Terminals A and B while removing all the independent sources.

You can see the Thévenin equivalent in Circuit F of Figure 8-8. The Thévenin equivalent reduces the source circuit of Circuit A to one voltage source in series with one resistor.

With the superposition method, I get the same expressions for v_{oc}, i_{sc}, and R_N as I get using source transformation (see the preceding section). You can use whichever method works best for you.

Gauging Maximum Power Transfer: A Practical Application of Both Theorems

The power p coming from the source circuit to be delivered to the load depends on both the current i flowing through the load circuit and the voltage v across the load circuit at the interface between the two circuits.

The *maximum power theorem* states that for a given source with a fixed Thévenin resistance R_T, the maximum power delivered to a load resistor R_L occurs when the R_L is matched or equal to R_T:

$$p_{max} \text{ when } R_L = R_T$$

Mathematically, the power is given by the following expression:

$$p = iv$$

The source circuit delivers maximum voltage when you have an open-circuit load. Because zero current flows through the open-circuit load, zero power is delivered to the load. Mathematically, the power p_{oc} delivered to the open-circuit load is

$$p_{oc} = iv = 0 \cdot v = 0$$

On the other hand, the source circuit delivers maximum current when you have a short-circuit load. Because zero voltage occurs across the short-circuit load, zero power is delivered to the load. Mathematically, the power p_{sc} delivered to the short-circuit load is

$$p_{sc} = iv = i \cdot 0 = 0$$

So what's the maximum power punch delivered for a given load resistance? Using either the Thévenin or Norton approach allows you to find the maximum power delivered to the load circuit.

To see how to determine the maximum power, look at the resistor arrangements for both the source and load circuits in Figure 8-9. In this figure, the source circuit is the Thévenin equivalent, and the load resistor is a simple but adjustable resistor.

Figure 8-9:
Determining
maximum
power
delivery.

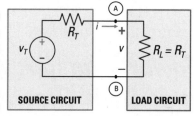

SOURCE CIRCUIT   LOAD CIRCUIT

Illustration by Wiley, Composition Services Graphics

Intuitively, you know that maximum power is delivered when both the current and voltage are maximized at the interface Terminals A and B.

Using voltage division (see Chapter 4), the voltage v across the interface at A and B is

$$v = \frac{R_L}{R_T + R_L} \cdot v_T$$

In Figure 8-9, the connected circuit between the source and load is a series circuit. The current i flows through each of the resistors, so

$$i = \frac{v_T}{R_T + R_L}$$

Substituting the values of v and i into the power equation, you wind up with the following power equation:

$$p = iv$$

$$p = \left(\frac{v_T}{R_T + R_L} \right)\left(\frac{R_L}{R_T + R_L} \cdot v_T \right)$$

$$p = \frac{R_L}{\left(R_T + R_L \right)^2} \cdot v_T^2$$

Determining the maximum power delivered to the load means taking the derivative of the preceding equation with respect to R_L and setting the derivative equal to zero. Here's the result:

$$\frac{dp}{dR_L} = \frac{R_T - R_L}{\left(R_T + R_L\right)^3} \cdot v_T^2 = 0$$

This equation equals zero when the numerator is zero. This occurs when $R_L = R_T$. Therefore, maximum power occurs when the source and load resistances are equal or matched.

Part III
Understanding Circuits with Transistors and Operational Amplifiers

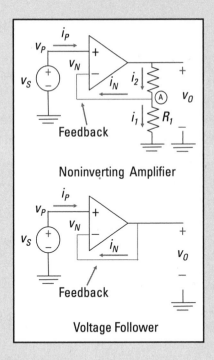

Noninverting Amplifier

Voltage Follower

Go to www.dummies.com/extras/circuitanalysis to solve a real-world problem on using a photoresistor and an op-amp circuit to convert light into a voltage with a desired output range.

In this part . . .

- ✔ Amplify current with transistors.
- ✔ Amplify voltage with operational amplifiers (also known as op amps).

Chapter 9

Dependent Sources and the Transistors That Involve Them

· ·

In This Chapter

▶ Working with linear dependent sources

▶ Analyzing circuits that have dependent sources

▶ Taking control with transistors

· ·

*R*esistors, capacitors, and inductors are interesting, but they're merely passive devices. What makes circuits great is the ability to perform as an electronic switch or amplify signals. Such switching and amplification functions are derived from transistors — *trans*fer re*sistors* — named for the fact that the resistance can be electronically tuned.

Most portable electronic devices, such as smartphones and tablets, use integrated circuits (ICs) to drive many system and circuit functions — making it possible for you to watch the latest YouTube sensation on the go. An IC is usually made on a small wafer of silicon or other semiconductor material holding hundreds to millions of transistors, resistors, and capacitors. In the future, gazillions of transistors, capacitors, and resistors could be jammed into a piece of silicon to perform other functions, like making coffee, getting your favorite newspaper, driving you to work, and waking you up to the reality of doing circuit analysis.

In this chapter, I introduce you to dependent sources, which you can use to model transistors and the operational amplifier IC, both of which require power to work. You analyze circuits with dependent sources using a variety of techniques from earlier chapters, and you explore some key types of dependent sources: JFET and bipolar transistors. As for operational amplifiers, I cover them in detail in Chapter 10.

Understanding Linear Dependent Sources: Who Controls What

A *dependent source* is a voltage or current source controlled by either a voltage or a current at the input side of the device model. The dependent source drives the output side of the circuit. Dependent sources are usually associated with components (or devices) requiring power to operate correctly. These components are considered *active devices* because they require power to work; circuits using these devices are called *active circuits*. Active devices such as transistors perform amplification, allowing you to do things like crank up the volume of your music.

When you're dealing with active devices operating in a linear mode, the relationship between the input and output behavior is directly proportional. That is, the bigger the input, the bigger the output. Mathematically for a given input *x,* you have an output *y* with a gain amplification of *G: y = Gx.*

The constant or gain *G* is greater than 1 for active circuits (think steroids) and less than 1 for passive circuits (think wimpy). In other engineering applications, technical terms for *G* include *scale factor, scalar multiplier, proportionality constant,* and *weight factor.*

The following sections introduce you to the four types of dependent sources and help you recognize the connection between dependent sources and their independent counterparts.

Classifying the types of dependent sources

Modeling active devices requires the use of dependent sources, and four types of dependent sources exist (see Figure 9-1):

- ✔ **Voltage-controlled voltage source (VCVS):** A voltage across the input terminals controls a dependent voltage source at the output port.

- ✔ **Current-controlled voltage source (CCVS):** A current flowing through the input terminals controls a dependent voltage source.

- ✔ **Voltage-controlled current source (VCCS):** Now account for a dependent current source at the output terminals. With a voltage across the input, you can control the amount of current output.

- ✔ **Current-controlled current source (CCCS):** Can you guess the last type of dependent source? That's right — with a current flowing through the input port, you can control a dependent current source at the output port.

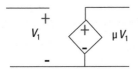

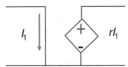

Voltage-controlled
voltage source (VCVS)

Current-controlled
voltage source (CCVS)

Figure 9-1:
Circuit
symbols of
dependent
sources.

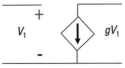

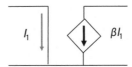

Voltage-controlled
current source (VCCS)

Current-controlled
current source (CCCS)

Illustration by Wiley, Composition Services Graphics

Diagrams use the diamond shape for dependent sources to distinguish them from independent sources, which use a circle. Some books may use circles to denote both dependent and independent sources. After a dependent source reaches 18 years old and leaves home, it becomes an independent source (just checking if you made it this far).

The output of a linear dependent source is proportional to the input voltage or current controlling the source output. In Figure 9-1, the proportionality constants or gains are given as μ, r, g, and β:

- ✔ You can think of μ in the VCVS dependent source as voltage gain because it's the ratio of the voltage output to the voltage input.

- ✔ In the CCVS dependent source, the proportionality constant r is called the *transresistance* because its input-output relationship takes the form of Ohm's law: $v = iR$.

- ✔ Similarly, the VCCS dependent source has a proportionality constant g, called the *transconductance,* following a variation of Ohm's law: $i = Gv$ (where the conductance $G = 1/R$).

- ✔ For the CCCS dependent source, you can think of the proportionality constant β as the current gain because it's the ratio of current output to current input.

So when the type of input matches the type of output, the proportionality constant gives you the current or voltage gain. When they differ, the input-output relationship stems from Ohm's law.

Recognizing the relationship between dependent and independent sources

You can turn off an independent voltage source by replacing it with a short circuit having zero resistance, and you can turn off an independent current source by replacing it with an open circuit — I show you how in Chapter 7. But you can't just turn dependent sources on or off. Because dependent sources rely on voltage or current from an independent source on the input side, turning off an independent source turns off a dependent source.

Figure 9-2 illustrates the interplay between the independent source and the dependent source. The top diagram shows that when an independent source is turned *on,* the dependent source is turned *on.* The bottom diagram shows that when an independent source is turned *off,* the dependent source is turned *off.*

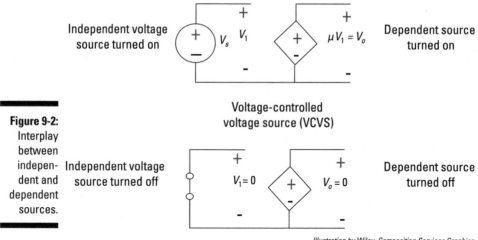

Figure 9-2: Interplay between independent and dependent sources.

Illustration by Wiley, Composition Services Graphics

Analyzing Circuits with Dependent Sources

This section shows how to analyze example circuits that have dependent sources by using the techniques I describe in Part II of this book. Get ready to use node-voltage analysis (Chapter 5), mesh-current analysis (Chapter 6), superposition (Chapter 7), and the Thévenin technique (Chapter 8) as you work with dependent sources. That's an impressive series of topics, so if you need an energy boost, feel free to grab a snack before you begin.

Applying node-voltage analysis

Using node voltage methods to analyze circuits with dependent sources follows much the same approach as for independent sources, which I cover in Chapter 5. Consider the circuit in Figure 9-3. What is the relationship between the output voltage v_o and i_s?

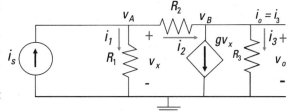

Illustration by Wiley, Composition Services Graphics

The first step is to label the nodes. Here, the bottom node is your reference node, and you have Node A (with voltage v_A) at the upper left and Node B (with voltage v_B) at the upper right. Now you can formulate the node voltage equations.

Using node-voltage analysis involves Kirchhoff's current law (KCL), which says the sum of the incoming currents is equal to the sum of the outgoing currents. At Node A, use KCL and substitute in the current expressions from Ohm's law ($i = v/R$). The voltage of each device is the difference in node voltages, so you get the following:

$$i_s = i_1 + i_2$$
$$i_s = \frac{v_A - 0}{R_1} + \frac{v_A - v_B}{R_2}$$

Rearranging gives you the node voltage equation:

$$\text{Node A: } \left(\frac{1}{R_1} + \frac{1}{R_2}\right)v_A + \left(-\frac{1}{R_2}\right)v_B = i_s$$

At Node B, again apply KCL and plug in the current expressions from Ohm's law:

$$i_2 = gv_x + i_3 \quad \left(\text{where } v_x = v_A - 0 = v_A\right)$$
$$\frac{v_A - v_B}{R_2} = gv_A + \frac{v_B - 0}{R_3}$$

Rearranging the preceding equation gives you the following node voltage equation at Node B:

$$\text{Node B: } \left(-\frac{1}{R_2}+g\right)v_A+\left(\frac{1}{R_3}+\frac{1}{R_2}\right)v_B=0$$

The two node voltage equations give you a system of linear equations. Put the node voltage equations in matrix form:

$$\begin{bmatrix} \frac{1}{R_1}+\frac{1}{R_2} & -\frac{1}{R_2} \\ -\frac{1}{R_2}+g & \frac{1}{R_3}+\frac{1}{R_2} \end{bmatrix}\begin{bmatrix} v_A \\ v_B \end{bmatrix}=\begin{bmatrix} i_s \\ 0 \end{bmatrix}$$

You can solve for the unknown node voltages v_A and v_B using matrix software. After you have the node voltages, you can set the output voltage v_o equal to v_B. You can then use the ever-faithful Ohm's law to find the output current i_o:

$$v_o=v_B-0=v_B \quad (\text{voltage across } R_3)$$

$$i_o=\frac{v_B}{R_3}=\frac{v_o}{R_3}$$

Using source transformation

To see the source transformation technique for circuits with dependent circuits, consider Circuit A in Figure 9-4. Suppose you want to find the voltage across resistor R_3. To do so, you can perform a source transformation, changing Circuit A (with an independent voltage source) to Circuit B (with an independent current source). You now have all the devices connected in parallel, including the dependent and independent current sources.

Don't use source transformation for dependent sources, because you may end up changing or losing the dependency. You need to make sure the dependent source is a function of the independent source.

Here's the equation for the voltage source and current source transformation:

$$i_s=\frac{v_s}{R_1}$$

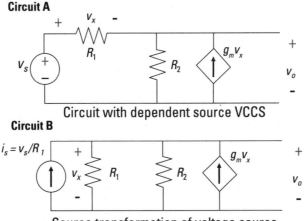

Circuit A

Circuit with dependent source VCCS

Circuit B

Source transformation of voltage source

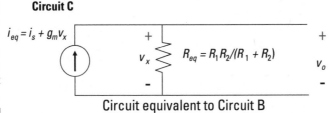

Figure 9-4:
Source
transformation for a
circuit with
a dependent
source.

Circuit C

Circuit equivalent to Circuit B

The independent current source i_s and the dependent current source gv_x point in the same direction, so you can add these two current sources to get the total current i_{eq} going through the resistor combination R_1 and R_2. The total current i_{eq} is $i_{eq} = i_s + g_m v_x$. Because v_x is the voltage across R_2, v_x is also equal to v_o in Circuit B: $v_o = v_x$.

Resistors R_1 and R_2 are connected in parallel, giving you an equivalent resistance R_{eq}:

$$R_{eq} = \frac{R_1 R_2}{R_1 + R_2}$$

The output voltage is equal to the voltage across R_{eq}, using Ohm's law and i_{eq}. You see the equivalent circuit with i_{eq} and R_{eq} in Circuit C. Because the dependent current source is dependent on v_x, you need to replace the voltage v_x with v_o:

$$v_o = i_{eq} R_{eq} = \left(i_s + g_m v_x \right) R_{eq}$$

$$v_o = \left(i_s + g_m v_o \right) \left(\frac{R_1 R_2}{R_1 + R_2} \right) \quad \text{(substitute } v_o = v_x \text{)}$$

Solving for the output voltage v_o gives you

$$v_o = i_s \left(\frac{R_{eq}}{1 - g_m R_{eq}} \right)$$

See how the output voltage is a function of the input source? The final expression of the output should not have a dependent variable.

Using the Thévenin technique

The Thévenin approach reduces a complex circuit to one with a single voltage source and a single resistor. Independent sources must be turned on because the dependent source relies on the excitation due to an independent source.

As I note in Chapter 8, to find the Thévenin equivalent for a circuit, you need to find the open-circuit voltage and the short-circuit current at the interface. In other words, you need to find the i-v relationship at the interface.

To see how to get the Thévenin equivalent for a circuit having a dependent source, look at Figure 9-5. This example shows how to find the input resistance and the output Thévenin equivalent circuit at interface points A and B.

Figure 9-5:
Thévenin
equivalent
for a
circuit with
a dependent
source.

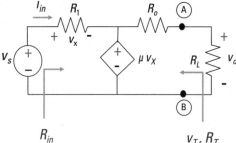

Illustration by Wiley, Composition Services Graphics

The input resistance is

$$R_{in} = \frac{v_s}{i_{in}}$$

Using Ohm's law, the current i_{in} through R_1 is

$$i_{in} = \frac{v_s - \mu v_x}{R_1}$$

$$i_{in} = \frac{v_s - \mu(i_{in}R_1)}{R_1}$$

Solving for i_{in}, you wind up with

$$i_{in} = \frac{v_s}{R_1(1+\mu)}$$

Substituting i_{in} into the input-resistance equation gives you

$$R_{in} = \frac{v_s}{i_{in}} = \frac{v_s}{\left(\dfrac{v_s}{R_1(1+\mu)}\right)}$$

$$R_{in} = R_1(1+\mu)$$

Here, the dependent source increases the input resistance by approximately multiplying the resistor R_1 by the dependent parameter μ. R_1 is the input resistance without the dependent source. To find the Thévenin voltage v_T and the Thévenin resistance R_T, you have to find the open-circuit voltage v_{oc} and short-circuit current i_{sc}. The resistance R_T is given by the following relationship:

$$R_T = \frac{v_{oc}}{i_{sc}}$$

Based on Figure 9-5, the open-circuit voltage is $v_{oc} = \mu v_x$. You find that the short-circuit current gives you

$$i_{sc} = \frac{\mu v_x}{R_o}$$

After finding v_{oc} and i_{sc}, you find the Thévenin resistance:

$$R_T = \frac{v_{oc}}{i_{sc}}$$

$$R_T = \frac{\mu v_x}{\left(\dfrac{\mu v_x}{R_o}\right)} = R_o$$

The output resistance R_o and Thévenin resistance R_T are equal. Based on Kirchhoff's voltage law (KVL), you have the following expression for v_x:

$$v_s = v_x + \mu v_x$$

$$v_x = \frac{v_s}{1 + \mu}$$

Substituting v_x into the equation for the open-circuit voltage v_{oc}, you wind up with

$$v_{oc} = \mu v_x$$

$$v_{oc} = v_s \left(\frac{\mu}{(1 + \mu)} \right)$$

The open-circuit voltage, v_{oc}, equals the Thévenin voltage, v_T. The nitty-gritty analysis leaves you with Thévenin voltage v_T and Thévenin resistance R_T, entailing a dependent voltage gain of μ:

$$v_T = v_s \left(\frac{\mu}{(1 + \mu)} \right)$$

$$R_T = R_o$$

When μ is very large, the Thévenin voltage v_T equals the source voltage v_s.

Describing a JFET Transistor with a Dependent Source

Transistors are amplifiers in which a small signal controls a larger signal. Just picture a Chihuahua taking its hefty owner for a daily walk, and you can imagine what a transistor is capable of.

The two primary types of transistors are bipolar transistors and field-effect transistors. The field-effect transistor is a little simpler than the bipolar kind. You can classify field-effect transistors, or FETs, in two ways: junction field-effect transistors (JFETs) and metal-oxide-semiconductor field-effect transistors (MOSFETs). Because JFETs provide a good picture of how transistor circuits work, this section concentrates on this type of FET. (I cover bipolar transistors in the later section "Examining the Three Personalities of Bipolar Transistors.")

Typical transistors have three leads. In the case of a JFET, a voltage on one lead (called the *gate*) is used to control a current between the two other leads (called the *source* and the *drain*). The gate voltage needs to be referenced to some other voltage, and by convention, it's referenced to the source terminal. Figure 9-6 shows the JFET symbol and its corresponding dependence model. The gate, drain, and source labels (*G, D,* and *S,* respectively) are normally omitted, but I include them here for reference.

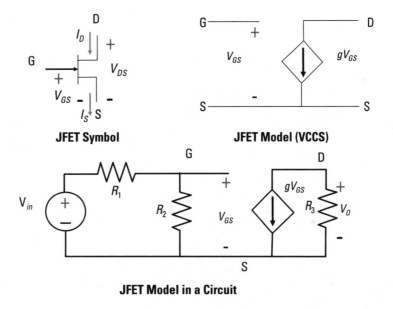

Figure 9-6:
JFET
transistor
symbol and
model and
the JFET
dependent
source
model in a
circuit.

JFET Symbol

JFET Model (VCCS)

JFET Model in a Circuit

Illustration by Wiley, Composition Services Graphics

In the figure, V_{GS} refers to the voltage between the gate and the source, I_D is the current into the drain, and I_S is the current out of the source. No current flows into the gate when it's operating under normal conditions, implying that the drain current I_D is equal to the source current I_S. A useful JFET model, which you see on the upper right of Figure 9-6, uses a voltage-controlled current source (VCCS). The model is part of the circuit at the bottom of the figure.

For the circuit in Figure 9-6, you need to find the ratio between the output voltage V_O and the input voltage V_{in}. The dependent source is a voltage-controlled current source, so its current is gV_{GS} (see the earlier section

"Classifying the types of dependent sources" for details). So at the output terminals of the dependent source model, the output voltage V_O is a result of the following equation using Ohm's law ($v = iR$):

$$V_O = \left(-gV_{GS}\right)R_3$$

The minus sign appears because the current through resistor R_3 flows in the opposite direction of the voltage polarities of the output voltage V_O. You can find the voltage V_{GS} on the input terminals of the dependent VCCS model.

Because the devices are connected in series on the input side of the circuit, you can use the voltage divider technique, as follows (see Chapter 4 for details on voltage division):

$$V_{GS} = V_{in}\left(\frac{R_2}{R_1 + R_2}\right)$$

Now substitute this expression for V_{GS} into the Ohm's law equation for output voltage V_O. You get the following input-output relationship:

$$V_O = \underbrace{V_{in}\left(\frac{R_2}{R_1 + R_2}\right)}_{=V_{GS}}\left(-gR_3\right)$$

$$V_O = \left(\frac{-gR_2R_3}{R_1 + R_2}\right)V_{in}$$

To see the amount of amplification using this circuit, try plugging in some numbers. Suppose $g = 1.8$ milliamps per volt, $R_1 = R_2 = 1$ kΩ, $R_3 = 10$ kΩ, and $V_{in} = 1$ volt. The amplifier output is

$$V_O = \left(\frac{-gR_2R_3}{R_1 + R_2}\right)V_{in}$$

$$V_O = \left(\frac{(-1.8 \text{ mA/V})(1 \text{ kΩ})(10 \text{ kΩ})}{1 \text{ kΩ} + 1 \text{ kΩ}}\right)(1 \text{ V}) = -9 \text{ V}$$

The input is amplified by –9 at the output of the dependent source. Awesome! The signal is bigger because an external voltage source made this JFET transistor work as an amplifier. The minus sign means that the signal is inverted or upside down, which is no problem because it doesn't change the sound quality of your music.

Examining the Three Personalities of Bipolar Transistors

Along with FETs (field-effect transistors — see the preceding section), bipolar transistors stand as a cornerstone of modern microelectronics. The word *bipolar* comes from the flow of both electrons and holes (where a *hole* is a positively charged particle). Because the bipolar transistor is a three-terminal device, the voltage between two terminals controls the current through the third terminal. The three terminals are called the *base,* the *emitter,* and the *collector.* Figure 9-7 shows the circuit symbols for two types of bipolar transistors: NPN and PNP. (For detailed information on working with transistors, check out *Electronics All-in-One For Dummies.*)

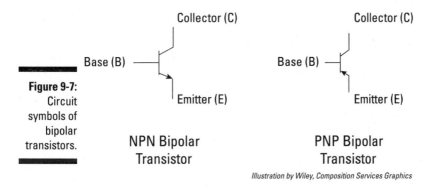

Figure 9-7:
Circuit
symbols of
bipolar
transistors.

Collector (C)   Collector (C)

Base (B)   Base (B)

Emitter (E)   Emitter (E)

NPN Bipolar   PNP Bipolar
Transistor   Transistor

Illustration by Wiley, Composition Services Graphics

Because of the versatility of the transistors, there are three basic patterns of design to perform circuit functions (see Figure 9-8 for the visual):

- **Common emitter:** *Common emitter* means that the emitter terminal is common to both the input and output parts of a circuit. The same holds true for the other two configurations described in this list.

- **Common base:** When the base terminal is common to both the input and output parts of the circuit, you have a common base circuit.

- **Common collector:** For the common collector arrangement, the collector terminal comes into play for both input and output circuit pieces.

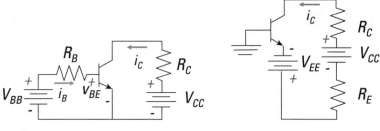

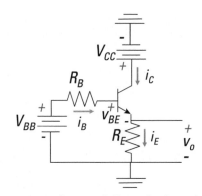

Illustration by Wiley, Composition Services Graphics

Figure 9-8:
Three
common
circuit con-
figurations
of bipolar
transistors.

You can use these circuits in stages or in combination to perform useful functions. That's what you do with an operational amplifier (op amp), which I
cover in the next chapter. I don't go through all the benefits of the transistor
configurations in the following sections, but I do give you a glimpse of the
device's worth.

Making signals louder with the common emitter circuit

A current-controlled current source (CCCS) is a typical model when you're
analyzing a circuit that has a bipolar transistor. Figure 9-9 shows a CCCS as
a very simplified DC model of a bipolar transistor. Note that γ denotes some
threshold on the voltage.

Figure 9-9:
Modified
current-
controlled
current
source for a
bipolar
transistor.

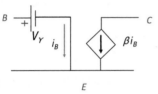

Illustration by Wiley, Composition Services Graphics

Figure 9-10 shows a common emitter circuit modeled with a dependent source. Note the labeling of the three terminals: base (B), collector (C), and emitter (E). Here, you use mesh-current analysis to find the transistor base current i_B.

Figure 9-10:
Simple
dependent
CCCS model
of a bipolar
transistor.

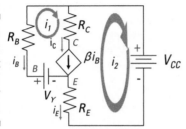

Illustration by Wiley, Composition Services Graphics

You see two mesh currents, i_1 and i_2, in Figure 9-10. By excluding the current source, you can combine Meshes 1 and 2 to create a supermesh. Going through the dependent current source isn't helpful for this analysis, and the supermesh is useful because it lets you avoid the series combination of dependent source βi_B and collector resistance R_C.

Starting at the bottom of the circuit, you can write the following KVL mesh expression:

$$i_2 R_E - V_\gamma + i_1 R_B + V_{CC} = 0 \qquad \text{KVL equation}$$

This equation has two unknown mesh currents, i_1 and i_2, through the dependent source βi_B. Use these concepts to write the KCL equation:

$$i_1 - i_2 = \beta i_B \qquad \text{KCL equation}$$

You use the KVL and KCL equations to solve for the unknown mesh currents. Figure 9-10 shows that the base current is in the opposite direction of the mesh current i_1:

$$i_B = -i_1$$

Substitute the value of i_B into the KCL equation to get a relationship between i_1 and i_2:

$$i_1 - i_2 = \beta i_B = -\beta i_1$$
$$i_2 = (\beta + 1)i_1$$

You now have current gain to amplify signals:

$$\frac{i_2}{i_1} = \beta + 1$$

The analysis shows the bipolar transistor as a current amplifier. Any change in i_1 (or the base current i_B) creates an even larger change in the collector or emitter current related by i_2. You need current amplifiers for current-hungry devices, such as speakers and magnetic locks to keep unsavory characters out of your house.

To find the input resistance at DC, solve the preceding equation for i_2 and substitute it into the KVL equation. Solve the KVL equation for i_1, which equals $-i_B$:

$$\underbrace{(\beta + 1)i_1}_{=i_2} R_E - V_\gamma + i_1 R_B + V_{CC} = 0$$

$$-i_1 = i_B = \frac{V_{CC} - V_\gamma}{R_B + (\beta + 1)R_E}$$

Here's the input resistance:

$$R_{in} = \frac{V_{CC} - V_\gamma}{-i_1}$$

$$R_{in} = \frac{(R_B + (\beta + 1)R_E)i_1}{i_1}$$

$$R_{in} = R_B + (\beta + 1)R_E$$

The dependent source increased the input resistance by adding an emitter resistor R_E and making R_E larger by about β. High-input resistance isolates the input and output parts of the circuit from from each other. Neato! Why neato? Because there aren't any loading effects. You design input circuits independently from output circuits. When you tie together these two circuits, they perform as expected. Each circuit is unaffected, but they work together to create the desired outcome — redesign isn't needed.

Amplifying signals with a common base circuit

The common base circuit looks like an ideal current source and is often called a *current buffer*. Figure 9-11 shows a hybrid-π model for AC signal analysis with infinite output resistance for an NPN transistor, which I use to analyze the common base configuration. The model has an input resistance r_π and a current gain β.

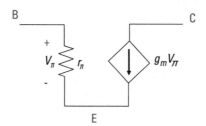

Hybrid-π model

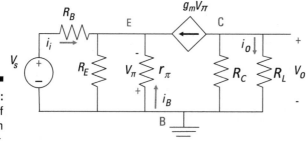

Figure 9-11:
Analysis of
a common
base circuit.

Small-signal equivalent

Illustration by Wiley, Composition Services Graphics

The circuit shows the parallel connection between the collector resistor R_C and load resistor R_L (symbolized by $R_C \| R_L$). Applying Ohm's law leads you to the small signal output voltage V_o:

$$V_o = -\left(g_m V_\pi\right)\left(R_C \| R_L\right)$$

At the emitter node, you apply KCL to get

$$g_m V_\pi + \frac{V_\pi}{r_\pi} + \frac{V_\pi}{R_E} + \frac{\left(V_s - \left(-V_\pi\right)\right)}{R_B} = 0 \qquad \text{(KCL)}$$

For the hybrid-π model, you have the transistor parameters related by the following equation, with β given earlier as the transistor current gain:

$$\beta = g_m r_\pi$$

Substituting β into the KCL equation, you wind up with

$$V_\pi\left(\frac{1+\beta}{r_\pi} + \frac{1}{R_E} + \frac{1}{R_B}\right) = -\frac{V_s}{R_B}$$

The terms in brackets form a parallel connection of the given resistors. Solving for V_π you have the following expression:

$$V_\pi = -\frac{V_s}{R_B}\left(\left(\frac{r_\pi}{1+\beta}\right)\| R_E \| R_B\right)$$

Substituting V_π into the V_o equation gives you the small-signal gain:

$$A_v = \frac{V_o}{V_s} = g_m\left(\frac{R_C \| R_L}{R_B}\right)\left(\left(\frac{r_\pi}{1+\beta}\right)\| R_E \| R_B\right)$$

$$A_v \approx g_m\left(R_C \| R_L\right) \quad \text{where } R_B \to 0, \text{ then } \left(\frac{r_\pi}{1+\beta}\right)\| R_E \| R_B \approx R_B$$

To find the current gain, you let the emitter resistance R_E approach infinity. This is reasonable because R_E is large compared to R_B and $r_\pi/(\beta + 1)$. This means little or no current will flow through R_E. Using $\beta = g_m r_\pi$, the KCL at the emitter node is

$$i_i + i_B = -g_m V_\pi = -\beta i_B$$
$$i_i = -\left(\beta + 1\right)i_B$$

The KCL equation at the collector node is

$$i_o = -g_m V_\pi = -\underbrace{g_m r_\pi}_{=\beta} i_B = -\beta i_B$$

Using the current divider equation at the output side, the current gain is defined as the ratio of i_o to i_i:

$$\frac{i_o}{i_i} = \left(\frac{\beta}{\beta+1}\right)\left(\frac{R_C}{R_C + R_L}\right)$$

The preceding equation shows the current gain is less than 1. There's some power gain because the voltage gain is greater than 1.

Isolating circuits with the common collector circuit

The common collector circuit is also known as an *emitter follower*. This means that any variation in the base terminal causes the same variation in the emitter terminal. However, the following analysis deals with signals that are constant, also known as *DC analysis*.

The following analysis of the circuit in Figure 9-12 shows that the emitter follower has a voltage gain that's approximately equal to 1 but provides a high input impedance, isolating the source circuit from the load circuit.

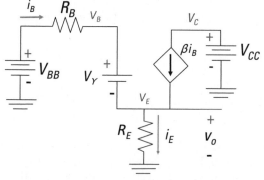

Figure 9-12:
A common
collector
circuit.

Common Collector Configuration

Illustration by Wiley, Composition Services Graphics

Using Figure 9-12, you get the output voltage with Ohm's law:

$$v_o = i_E R_E$$

Apply KVL to get the base voltage v_B:

$$v_B = V_\gamma + v_o$$

Because V_γ is a constant (about 0.7 volts for a silicon-based transistor and 0.2 volts for a germanium-based transistor), the output voltage v_o follows the input voltage v_B. So the common collector doesn't produce a voltage gain, but it does provide circuit isolation to reduce circuit loading due to its high input impedance.

Find the input resistance as seen by the base terminal as follows:

$$R_{in} = \frac{v_B}{i_B}$$

The output voltage v_o follows from Ohm's law:

$$v_o = R_E i_E$$

But the emitter current i_E is related to the input base current i_B:

$$i_E = (\beta + 1) i_B$$

You now have an output voltage:

$$v_o = R_E i_E = R_E (\beta + 1) i_B$$

If the output voltage v_o is much bigger than V_γ, you can make the following approximation for the base voltage v_B:

$$v_B = V_\gamma + v_O \approx v_O$$
$$v_B = R_E (\beta + 1) i_B$$

Solving for the ratio of v_B to i_B gives you the input resistance:

$$R_{in} = \frac{v_B}{i_B} = R_E (\beta + 1)$$

The input resistance as seen by the base multiplies the emitter resistance by $\beta + 1$. Typical values of current gain β vary from 50 to 150. High-input resistance provides isolation between the input and output parts of the circuit. Also, when little current is drawn from the source, you have longer battery life for portable applications, letting you play games on your smartphone for longer stretches.

Chapter 10

Letting Operational Amplifiers Do the Tough Math Fast

. . . .

In This Chapter

▶ Performing hand calculations electronically with operational amplifiers

▶ Modeling secrets of operational amplifiers with dependent sources

▶ Taking a look at op-amp circuits

▶ Putting together some systems

. . . .

*T*he operational amplifier (op amp) is a powerful tool when you're working with active devices in modern-day circuit applications. Because op amps can do calculations electronically, they perform mathematical operations (like addition, subtraction, multiplication, division, integration, and derivatives) fast. You can put together basic op-amp circuits to build accurate mathematical models that predict complex and real-world behavior — like when the breakfast pastry in your toaster will turn into a flaming torch.

This chapter introduces op-amp circuits, demonstrates how to use them to perform certain mathematical operations, and gives you a peek at the more complex processing actions that op amps serve as the building blocks for.

The Ins and Outs of Op-Amp Circuits

Commercial op amps first entered the market as integrated circuits in the mid-1960s, and by the early 1970s, they dominated the active device market in analog circuits. The op amp itself consists of a complex arrangement of transistors, diodes, resistors, and capacitors put together and built on a tiny silicon chip called an *integrated circuit*.

You can model the op amp with simple equations with little concern for what's going on inside the chip. You just need some basic knowledge of the constraints on the voltages and currents at the external terminals of the device.

In the following sections, you discover typical diagrams of op-amp circuits, the characteristics of ideal op amps and op amps with dependent sources, and the two equations necessary for analyzing these special circuits.

Discovering how to draw op amps

Unlike capacitors, inductors, and resistors, op amps require power to work. Op amps have the following five key terminals (see their symbols in Figure 10-1):

- The positive terminal, called the noninverting input v_P

- The negative terminal, called the inverting input v_N

- The output terminal, resulting from the voltage applied between noninverting and inverting inputs: $v_O = A(v_P - v_N)$

- Positive and negative power supply terminals, usually labeled as $+V_{CC}$ and $-V_{CC}$ and required for the op amp to operate correctly

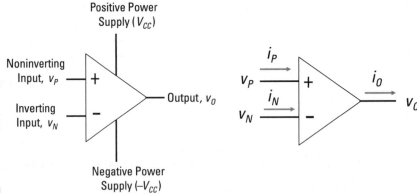

Figure 10-1: The circuit symbol of the op amp and its five terminals.

Illustration by Wiley, Composition Services Graphics

Although many op amps have more than five terminals, those terminals aren't normally shown symbolically. Also, to reduce the clutter when you're investigating an op-amp circuit, the power supplies aren't usually shown, either.

When the power supplies aren't shown in a diagram of an op-amp circuit, don't forget that the power supplies provide upper and lower limits of the output voltage, restricting its voltage range. Barring otherworldly powers, you can't get more power output than you supply.

Looking at the ideal op amp and its transfer characteristics

You can model the op amp with a dependent source if you need accurate results, but the ideal op amp is good enough for most applications.

The op amp amplifies the difference between the two inputs, v_P and v_N, by a gain A to give you a voltage output v_O:

$$v_O = A(v_P - v_N)$$

The voltage gain A for an op amp is very large — greater than 10^5.

When the output voltage exceeds the supplied power, the op-amp *saturates*. This means that the output is clipped or maxed out at the supplied voltages and can increase no further. When this happens, the op-amp behavior is no longer linear but operates in the nonlinear region.

You can see this idea in Figure 10-2. The left diagram shows the transfer characteristic, whereas the right diagram shows the ideal transfer characteristic of an op amp with an infinite gain. The graph shows three modes of operation for the op amp. You have positive and negative saturated regions, showing the nonlinear and linear regions. If you want to make signals bigger, you need to operate in the linear region. You can describe the three regions mathematically as follows

Negative saturated region:	$v_O = -V_{CC}$	$A_v(v_p - v_N) < -V_{CC}$
Linear active region:	$v_O = A_v(v_p - v_N)$	$-V_{CC} < A_v(v_p - v_N) < +V_{CC}$
Positive saturated region:	$v_O = +V_{CC}$	$A_v(v_p - v_N) > +V_{CC}$

To perform math functions (such as addition and subtraction), the op amp must work in linear mode. All op-amp circuits given in this chapter operate in the linear active region.

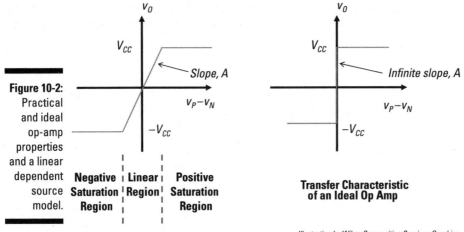

Figure 10-2: Practical and ideal op-amp properties and a linear dependent source model.

Modeling an op amp with a dependent source

If you need accurate results, you can model the op amp with a voltage-controlled dependent source, like the one in Figure 10-3. This model consists of a large gain A, a large input resistance R_I, and a small output resistance R_O. The table in Figure 10-3 shows ideal and typical values of these op-amp properties.

High amplification (or gain) makes the analysis simpler, allowing you not to worry about what's going on inside the op amp. As long as the op amp has high gain, the op-amp math circuits will work. High-input resistance draws little current from the input source circuit, increasing battery life for portable applications. Low- or no-output resistance delivers maximum voltage to the output load.

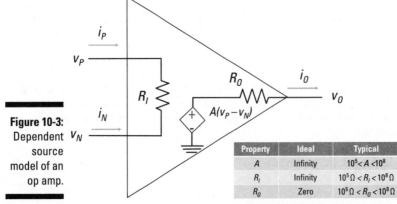

Figure 10-3: Dependent source model of an op amp.

Property	Ideal	Typical
A	Infinity	$10^5 < A < 10^8$
R_I	Infinity	$10^5\,\Omega < R_I < 10^8\,\Omega$
R_O	Zero	$10^5\,\Omega < R_O < 10^8\,\Omega$

The dependent voltage-controlled current source is shown in Figure 10-3. The output is restricted between the positive and negative voltages when the op amp is operating in the linear region.

Examining the essential equations for analyzing ideal op-amp circuits

The ideal properties of an op amp produce two important equations:

$$v_P = v_N$$
$$i_P = i_N = 0$$

These equations make analyzing op amps a snap and provide you with valuable insight into circuit behavior. Why? Because feedback from the output terminals to one or both inputs ensures that v_P and v_N are equal.

To get the first constraint, consider that the linear region of an op amp is governed by when the output is restricted by the supply voltages as follows:

$$-V_{CC} \le \underbrace{v_O}_{v_O = A(v_P - v_N)} \le +V_{CC}$$

You can rearrange the equation to limit the input to $v_P - v_N$:

$$\frac{-V_{CC}}{A} \le v_P - v_N \le \frac{+V_{CC}}{A}$$

For an ideal op amp, the gain A is infinity, so the inequality becomes

$$0 \le v_P - v_N \le 0$$

Therefore, the ideal op amp (with infinite gain) must have this constraint:

$$v_P = v_N$$

An op amp with infinite gain will always have the noninverting and inverting voltages equal. This equation becomes useful when you analyze a number of op-amp circuits, such as the op-amp noninverter, inverter, summer, and subtractor.

The other important op amp equation takes a look at the input resistance R_I. An ideal op amp has infinite resistance. This implies that no input currents can enter the op amp:

$$i_P = i_N = 0$$

The equation says that the op-amp input terminals act as open circuits.

You need to connect the output terminal to the inverting terminal to provide negative feedback in order to make the op amp work. If you connect the output to the positive side, you're providing positive feedback, which isn't good for linear operation. With positive feedback, the op amp would either saturate or cause its output to undergo oscillations.

Looking at Op-Amp Circuits

The mathematical uses for signal processing include noninverting and inverting amplification, addition, and subtraction. Doing these math operations simply requires resistors and op amps. (As a gentle reminder to some students who are troubleshooting their circuits, you also need power.) In the following sections, you analyze several op-amp configurations using the op-amp equations: $v_P = v_N$ and $i_P = i_N = 0$.

Analyzing a noninverting op amp

One of the most important signal-processing applications of op amps is to make weak signals louder and bigger. The following example shows how the feedback affects the input-output behavior of an op-amp circuit. Consider Figure 10-4, which first shows the input connected to the noninverting input. You have a feedback path from the output circuit leading to the inverting input.

The voltage source v_S connects to the noninverting input v_P:

$$v_P = v_S$$

You gotta first find the voltage at the inverting input so you can figure out how the input and output voltages are related. Apply Kirchhoff's current law (KCL) at Node A between resistors R_1 and R_2. (Remember that KCL says the sum of incoming currents is equal to the outgoing currents.) Applying KCL gives you

$$i_2 = i_N + i_1$$

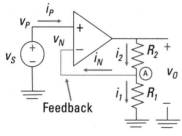
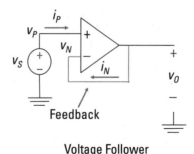

Figure 10-4: Noninverting amplifier op amp and voltage follower.

Noninverting Amplifier   Voltage Follower

Illustration by Wiley, Composition Services Graphics

At the output side of the op amp, the inverting current i_N is equal to zero because you have infinite resistance at the inverting input. This means that all the current going through resistor R_2 must go through resistor R_1. If the current is the same, R_1 and R_2 must be connected in series, giving you

$$i_2 = i_1$$

Because resistors R_1 and R_2 are connected in series, you can use voltage division (see Chapter 4 for details). Voltage division gives you the voltage relationship between the inverting input v_N and output v_O:

$$v_N = \left(\frac{R_1}{R_1 + R_2} \right) v_O$$

The inverting input v_N and noninverting input v_P are equal for ideal op amps. So here's the link between the input source voltage v_S and output voltage v_O:

$$v_P = v_N \quad \rightarrow \quad v_S = \left(\frac{R_1}{R_1 + R_2} \right) v_O$$

You now have the ratio of the voltage output to the input source:

$$\frac{v_O}{v_S} = \frac{R_1 + R_2}{R_1}$$

$$\frac{v_O}{v_S} = 1 + \frac{R_2}{R_1}$$

You just made the input voltage v_S larger by making sure the ratio of the two resistors is greater than 1. You read right: To make the input signal louder, the feedback resistor R_2 should have a larger value than input resistor R_1.

Piece of cake! For example, if $R_2 = 9$ kΩ and $R_1 = 1$ kΩ, then you have the following output voltage:

$$v_O = \left(1 + \frac{9 \text{ k}\Omega}{1 \text{ k}\Omega}\right)v_S$$
$$v_O = 10v_S$$

You've amplified the input voltage by ten. Awesome!

Following the leader with the voltage follower

A special case of the noninverting amplifier is the *voltage follower*, in which the output voltage follows in lock step with whatever the input signal is. In the voltage follower (pictured in Figure 10-4), v_S is connected to the noninverting terminal. You can express this idea as

$$v_P = v_S$$

You also see that the output v_O is connected to the inverting terminal, so

$$v_N = v_O$$

An ideal op amp has equal noninverting and inverting voltage. This means that the preceding two equations are equal. In other words

$$v_O = v_S$$

You can also view the voltage follower as a special case of the noninverting amplifier with a gain of 1, because the feedback resistor R_2 is zero (a short circuit) and resistor R_1 is infinite (open circuit):

$$\frac{v_O}{v_S} = 1 + \frac{\overset{0}{\overbrace{R_2}}}{\underset{\infty}{\underbrace{R_1}}} = 1$$

$$v_O = v_S$$

The output voltage v_O is equal to the input source voltage v_S. The voltage gain is 1 where the output voltage follows the input voltage. But a piece of wire gives a gain of 1, too, so what good is this circuit? Well, the voltage follower provides a way to put together two separate circuits without having them affect each other. When they do affect each other in a bad way, that's called *loading.* A voltage follower solves the loading problem.

Turning things around with the inverting amplifier

An inverting amplifier takes an input signal and turns it upside down at the op-amp output. When the value of the input signal is positive, the output of the inverting amplifier is negative, and vice versa.

Figure 10-5 shows an inverting op amp. The op amp has a feedback resistor R_2 and an input resistor R_1 with one end connected to the voltage source. The other end of the input resistor is connected to the inverting terminal, and the noninverting terminal is grounded at 0 volts. The amount of amplification depends on the ratio between the feedback and input resistor values.

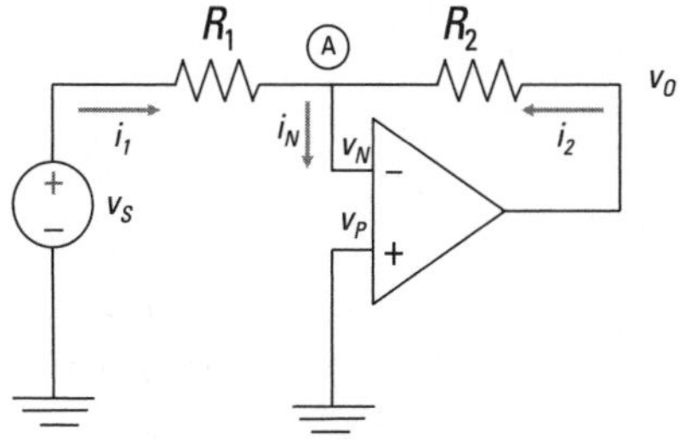

Figure 10-5:
A standard
inverting
amplifier.

Illustration by Wiley, Composition Services Graphics

Because the noninverting input is grounded to 0 volts, you have

$$v_P = v_N = 0$$

For ideal op amps, the voltages at the inverting and noninverting terminals are equal and set to zero. The inverting terminal is connected to a virtual ground because it's indirectly connected to ground by v_P.

Apply Kirchhoff's current law (KCL) to Node A to get the following:

$$\frac{v_S - v_N}{R_1} + \frac{v_O - v_N}{R_2} - i_N = 0 \quad (\text{KCL})$$

Simplify the equation with the following constraints for an ideal op amp:

$$v_P = v_N = 0$$
$$i_P = i_N = 0$$

The constraints make the KCL equation simpler:

$$\frac{v_S}{R_1} + \frac{v_O}{R_2} = 0$$

You wind up with the following relationship between the input and output voltages:

$$\frac{v_O}{v_S} = -\frac{R_2}{R_1}$$

Again, the amplification of the signal depends on the ratio of the feedback resistor R_2 and input resistor R_1. You need only external components of the op amp to make the signal way bigger. The negative sign means the output voltage is an amplified but inverted (or upside down) version of the input signal.

For a numerical example, let R_2 = 10 kΩ and R_1 = 1 kΩ. In that case, the inverted output voltage v_O is ten times as big as the input voltage v_S. In nothing flat, you just made a weak signal stronger. Nice work — you deserve a raise!

Resistors should be around the 1 kΩ to 100 kΩ range to minimize the effects of variation in the op-amp characteristics and voltage sources.

Adding it all up with the summer

You can extend the inverting amplifier to more than one input to form a *summer,* or *summing amplifier.* Figure 10-6 shows an inverting op amp with two inputs. The two inputs connected at Node A (called a *summing point*) are connected to an inverting terminal.

Because the noninverting input is grounded, Node A is also connected as a virtual ground. Applying the KCL equation at Node A, you wind up with

$$i_1 + i_2 + i_F - i_N = 0$$

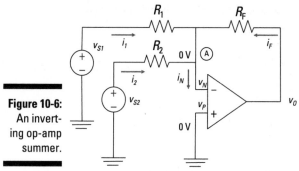

Figure 10-6:
An inverting op-amp summer.

Illustration by Wiley, Composition Services Graphics

Replace the input currents in the KCL equation with node voltages and Ohm's law ($i = v/R$):

$$\frac{v_{S1}-0}{R_1}+\frac{v_{S2}-0}{R_2}+\frac{v_O-0}{R_F}-\underset{0}{i_N}=0$$

Because $i_N = 0$ for an ideal op amp, you can solve for the output voltage in terms of the input source voltages:

$$v_O =\underbrace{\left(-\frac{R_F}{R_1}\right)}_{G_1}v_{S1}+\underbrace{\left(-\frac{R_F}{R_2}\right)}_{G_2}v_{S2}$$

$$v_O = G_1 v_{S1} + G_2 v_{S2}$$

The output voltage is a weighted sum of the two input voltages. The ratios of the feedback resistance to the input resistances determine the gains, G_1 and G_2, for this op-amp configuration.

To form a summing amplifier (or inverting summer), you need to set the input resistors equal with the following constraint:

$$R_1 = R_2 = R$$

Applying this constraint gives you the output voltage:

$$v_O =\left(-\frac{R_F}{R}\right)(v_{S1}+v_{S2})$$

This shows that the output is proportional to the sum of the two inputs. You can easily extend the summer to more than two inputs.

Plug in the following values for Figure 10-6 to test this mumbo jumbo: v_{S1} = 0.7 volts, v_2 = 0.3 volts, R_1 = 7 kΩ, R_2 = 3 kΩ, and R_F = 21 kΩ. Then calculate the output voltage v_o:

$$v_O = \left(-\frac{R_F}{R_1}\right)v_{S1} + \left(-\frac{R_F}{R_2}\right)v_{S2}$$

$$v_O = \left(-\frac{21 \text{ k}\Omega}{7 \text{ k}\Omega}\right)(0.7 \text{ V}) + \left(-\frac{21 \text{ k}\Omega}{3 \text{ k}\Omega}\right)(0.3 \text{ V})$$

$$v_O = -4.2 \text{ V}$$

The signals are bigger — mission accomplished. If signals are changing in time, the summer adds these signals instantly with no problem.

What's the difference? Using the op-amp subtractor

You can view the next op-amp circuit — which is a *differential amplifier,* or *subtractor* — as a combination of a noninverting amplifier and inverting amplifier (see the earlier related sections for the scoop on these amplifiers). Figure 10-7 shows an op-amp subtractor.

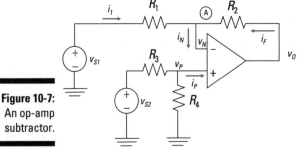

Figure 10-7: An op-amp subtractor.

Illustration by Wiley, Composition Services Graphics

You use superposition to determine the input and the output relationship. As I explain in Chapter 7, the superposition technique involves the following steps:

1. **Turn on one source and turn off the others.**

2. **Determine the output of the source that's on.**

3. **Repeat for each input, taking the sources one at a time.**

4. **Algebraically add up all the output contributions for each input to get the total output.**

For Figure 10-7, first turn off voltage source v_{S2} so that there's no input at the noninverting terminal ($v_P = 0$). With the noninverting input grounded, the circuit acts like an inverting amplifier. You wind up with output contribution v_{O1} due to v_{S1}:

$$v_{O1} = \left(-\frac{R_2}{R_1} \right) v_{S1}$$

You next turn off voltage source v_{S1} ($v_N = 0$) and turn v_{S2} back on. The circuit now acts like a noninverting amplifier. Because this is an ideal op amp, no current ($i_P = 0$) is drawn from the series connection of resistors R_3 and R_4, so you can use the voltage divider equation to determine v_P. The voltage at the noninverting input is given by

$$v_P = \left(\frac{R_4}{R_3 + R_4} \right) v_{S2}$$

The noninverting input v_P is amplified to give you an output v_{O2}:

$$v_{O2} = \left(1 + \frac{R_2}{R_1} \right) v_P$$

$$v_{O2} = \left(1 + \frac{R_2}{R_1} \right) \left(\frac{R_4}{R_3 + R_4} \right) v_{S2}$$

You then add up the outputs v_{O1} and v_{O2} to get the total output voltage:

$$v_O = v_{O1} + v_{O2}$$

$$= -\underbrace{\left(\frac{R_2}{R_1} \right)}_{G_1} v_{S1} + \underbrace{\left(1 + \frac{R_2}{R_1} \right) \left(\frac{R_4}{R_3 + R_4} \right)}_{G_2} v_{S2}$$

$$= -G_1 v_{S1} + G_2 v_{S2}$$

Here, $-G_1$ is the inverting gain and G_2 is the noninverting gain. You need the following constraint to form a subtractor:

$$\frac{R_3}{R_1} = \frac{R_4}{R_2}$$

Applying the constraint simplifies the output voltage, giving you

$$v_O = \frac{R_2}{R_1}\left(v_{S2} - v_{S1}\right)$$

There you have it! You now have the output proportional to the difference between the two inputs.

Increasing the Complexity of What You Can Do with Op Amps

If you've already read the earlier section "Looking at Op-Amp Circuits," then you have the basic building blocks of op-amp circuits and are ready to tackle the complex processing actions I describe next.

Analyzing the instrumentation amplifier

The instrumentation amplifier is a differential amplifier suited for measurement and test equipment. Figure 10-8 shows the input stage of an instrumentation amplifier. Your goal is to find the voltage output v_O proportional to the difference of the two inputs, v_1 and v_2. Getting the desired output requires some algebraic gymnastics, but you can handle it.

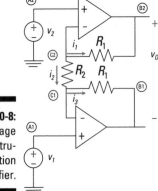

Figure 10-8:
Input stage
of an instru-
mentation
amplifier.

Illustration by Wiley,
Composition Services Graphics

At Node C2, you apply KCL ($i_1 + i_2 = 0$) and Ohm's law ($i = v/R$) and wind up with

Node C2: $\dfrac{v_{C2} - v_{B2}}{R_1} + \dfrac{v_{C2} - v_{C1}}{R_2} = 0$

At Node C1, the KCL equation ($-i_2 + i_3 = 0$) with Ohm's law leads you to

Node C1: $\dfrac{v_{C1} - v_{C2}}{R_2} + \dfrac{v_{C1} - v_{B1}}{R_1} = 0$

Figure 10-8 shows the noninverting input connected to independent voltages v_1 and v_2. Use the op-amp voltage constraint $v_P = v_N$ to get the following:

$v_{C2} = v_2$ and $v_{C1} = v_1$

Substitute v_1 and v_2 into KCL equations, which gives you

Node C2: $\dfrac{v_2 - v_{B2}}{R_1} + \dfrac{v_2 - v_1}{R_2} = 0$

Node C1: $\dfrac{v_1 - v_2}{R_2} + \dfrac{v_1 - v_{B1}}{R_1} = 0$

Now solve for v_{B2} and v_{B1}, because the output voltage v_O depends on these two values:

$v_{B2} = v_2 + \dfrac{R_1}{R_2}(v_2 - v_1)$

$v_{B1} = v_1 + \dfrac{R_1}{R_2}(v_1 - v_2)$

The output voltage v_O is the difference between the v_{B1} and v_{B2}:

$v_O = v_{B2} - v_{B1}$

$= \left[v_2 + \dfrac{R_1}{R_2}(v_2 - v_1)\right] - \left[v_1 + \dfrac{R_1}{R_2}(v_1 - v_2)\right]$

$= \left(1 + \dfrac{2R_1}{R_2}\right)(v_2 - v_1)$

Cool! Resistor R_2 can be used to amplify the difference $v_2 - v_1$. After all, it's easier to change the value of one resistor R_2 than of two resistors R_1.

Implementing mathematical equations electronically

As an example of how op amps can solve equations, consider a single output and three voltage input signals:

$$v_O = 10v_1 + 5v_2 - 4v_3$$

You can rewrite the equation in many ways to determine which op-amp circuits you need to perform the math. Here's one way:

$$v_O = -10(-v_1) - 5(-v_2) - 4(v_3)$$

The equation suggests that you have an inverting summer with three inputs: $-v_1$, $-v_2$, and v_3. You need an inverting amplifier with a gain of -1 for v_1 and v_2. Input v_1 has a summing gain of -10, input v_2 has a summing gain of -5, and input v_3 has a summing gain of -4. You can see one of many possible op-amp circuits in the top diagram of Figure 10-9. The dashed boxes indicate the two inverting amplifiers and the inverting summer.

The outputs of the two inverting amplifiers are $-v_1$ and $-v_2$, and they're inputs to the inverting summer. The third input to the summer is v_3. Adding up the three inputs with required gains entails an inverting summer, which you see in Figure 10-9.

For input v_1, the ratio of the inverting summer's feedback resistor of 200 kΩ to its input resistor of 20 kΩ provides a gain of -10. Similarly, for input v_2, the ratio of the feedback resistor of 200 kΩ to its input resistor of 40 kΩ gives you a gain of -5. Finally, for input v_3, the ratio of the feedback resistor of 200 kΩ to its input resistor of 50 kΩ provides a gain of -4. You can use other possible resistor values as long as the ratio of resistors provides the correct gains for each input.

Reducing the number of op amps during the design process helps lower costs. And with some creativity, you can reduce the number of op amps in the circuit by rewriting the math equation of the input-output relationship:

$$v_O = 10v_1 + 5v_2 - 4v_3$$
$$v_O = -2\left[-5v_1 - 2.5v_2\right] - 4v_3$$

Part IV

Applying Time-Varying Signals to First- and Second-Order Circuits

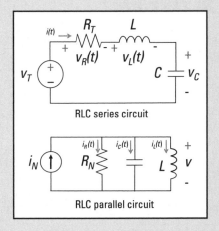

RLC series circuit

RLC parallel circuit

Explore a timing circuit that detects rectangular pulses at www.dummies.com/
extras/circuitanalysis.

In this part . . .

✔ Look at functions that describe AC signals, such as the step function and the exponential function.

✔ Get acquainted with capacitors and inductors and the roles they play in circuits.

✔ Find out how to analyze first-order circuits (circuits with a single storage element connected to a single resistor or a resistor network).

✔ Practice analyzing second-order circuits, which consist of capacitors, inductors, and resistors.

Chapter 11

Making Waves with Funky Functions

In This Chapter

▶ Observing spikes with the impulse function

▶ Creating step functions

▶ Rising or falling with the exponential function

▶ Cycling with sinusoidal functions

DC signals don't change with time . . . kinda boring, right? More interesting signals change in time like music. Such signals may spike, jump around, or rise and fall. They may build or decline steadily, or they may shoot up or plummet, picking up speed. They may repeat in cycles, continuing on and on.

Electric signals that change in time are useful because they can carry information about the real world, like temperature, pressure, and sound. This chapter covers basic time-varying signals commonly found in circuit analysis, including info on their key properties.

A word of warning: This chapter doesn't meet the high benchmarks of the Grand Poobah of precision math, but it's good enough to play with some funky functions.

Spiking It Up with the Lean, Mean Impulse Function

The first funky function is one you may have never heard of, but it occurs frequently in real life. It's called an *impulse function*, also known as a *Dirac delta function*. Just think of the impulse as a single spike that occurs in one instant of time. You can view this spiked function as one that's infinitely large in magnitude and infinitely thin in time, having a total area of 1.

You can visualize the impulse as a limiting form of a rectangular pulse of unit area. Specifically, as you decrease the duration of the pulse, its amplitude increases so that the area remains constant at unity. The more you decrease the duration, the closer the rectangular pulse comes to the impulse function. The bottom diagram of Figure 11-1 shows the limiting form of the rectangular pulse approaching an impulse. (Check out the nearby sidebar "Identifying impulse functions in the day-to-day" if you're having trouble wrapping your head around impulse functions.)

So what's the practical use of the impulse function? By using the impulse as an input signal to a system, you can reveal the output behavior or character of a system. After you know the behavior of the system for an impulse, you can describe the system's output behavior for any input. Why is that? Because any input is modeled as a series of impulses shifted in time with varying heights, amplitudes, or strengths.

Here's the fancy pants description of the impulse function:

$$\delta(t) = 1 \qquad t = 0$$
$$\delta(t) = 0 \qquad t \neq 0$$

Identifying impulse functions in the day-to-day

Some physical phenomena come very close to being modeled with impulse functions. One example is lightning. Lightning has lots of energy and occurs in a short amount of time. That fits the description of an impulse function. An ideal impulse has an infinitely high amplitude (high energy) and is infinitely thin in time. As you drive through a lightning storm, you may hear a popping noise if you're tuned in to a radio weather station. This noise occurs when the energy of the lightning interferes with the signal coming from the radio weather station.

Another example of a real-world impulse function is a bomb. A powerful bomb has lots of energy occurring in a short amount of time. Similarly, fireworks, including cherry bombs, produce loud noises — audio energy — that occur as a series of popping noises having short durations.

This mathematical description says that the impulse function occurs at only one point in time; the function is zero elsewhere. The impulse here occurs at the origin of time — that is, when you decide to let $t = 0$ (not at the beginning of the universe or anything like that).

The top-left diagram of Figure 11-1 shows an ideal unit impulse function having a large amplitude with a short duration. You can describe the area of the impulse function as the strength of the impulse:

$$\int_{-\infty}^{t} \delta(t)dx = 1u(t)$$

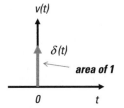

Ideal Impulse at $t = 0$

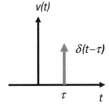

Delayed Ideal Impulse at $\tau = 0$

Figure 11-1:
The impulse
function,
delayed
impulse
function,
and rectan-
gular pulse.

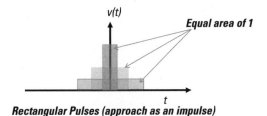

Rectangular Pulses (approach as an impulse)

Illustration by Wiley, Composition Services Graphics

At time $t = 0$, the area is a constant having a value of 1; and before $t = 0$, the area is equal to 0. The integration of the impulse results in another funky function, $u(t)$, called a *step function,* which I cover in the later section "Stepping It Up with a Step Function." You can view the impulse as a deriva-tive of the step function $u(t)$ with respect to time:

$$\delta(t) = \frac{d}{dt}\big[u(t)\big]$$

What these two equations tell you is that if you know one function, you can determine the other function.

In the following sections, I tell you how to change the strength of the impulse, delay the impulse, and evaluate an integral with an impulse function.

Changing the strength of the impulse

Figure 11-1 shows an impulse with an area (or strength) equal to 1. To have a different area or strength K, you can modify the impulse:

$$v(t) = K\delta(t)$$
$$\int_{-\infty}^{t} v(x)\, dx = Ku(t)$$

The area under the curve is given by strength K. The result of integrating the impulse leads you to another step function with amplitude or strength K.

Delaying an impulse

Impulses can be delayed. Analytically, you can describe a delayed impulse that occurs later, say, at time τ:

$$\delta(t-\tau) = 1 \quad t = \tau$$

$$\delta(t-\tau) = 0 \quad t \neq \tau$$

This equation says the impulse occurs only at a time later τ and nowhere else, or it's equal to 0 at time not equal to τ. You see a delayed impulse in the top-right diagram of Figure 11-1.

For a numerical example, let an impulse having a strength of 10 occur at delayed time $\tau = 5$. You can describe the delayed impulse as

$$10\delta(t-5) = 10 \quad t = \tau = 5$$
$$10\delta(t-5) = 0 \quad t \neq 5$$

The equation says that the impulse, which has strength $K = 10$, occurs only at a time $\tau = 5$ later and that the impulse occurs nowhere else. In other words, the impulse is equal to 0 when time is not equal to 5.

Evaluating impulse functions with integrals

Assuming $x(t)$ is a continuous function that's multiplied by a time-shifted (or delayed) impulse, the integral of the product is expressed and evaluated as follows:

$$\int_{-\infty}^{\infty} x(t)\delta(t-t_0)\,dt = x(t_0)u(t-t_0)$$

You do this evaluation only where the impulse occurs — at only one point and nowhere else. The preceding equation sifts out or selects the value of $x(t)$ at time equal to t_0. This integration is one of the easiest integrations you'll encounter.

Here's a simple numerical example with $x(t) = 5t^2 + 3t + 6$ and $t_0 = 5$:

$$x(5) = \int_{-\infty}^{\infty} \underbrace{\left[5t^2 + 3t + 6\right]}_{x(t)} \delta(t-\underset{t_0}{5})\,dt$$
$$= \left[5(5)^2 + 3(5) + 6\right]u(t-5)$$
$$= \left[125 + 15 + 6\right]u(t-5)$$
$$= 146u(t-5)$$

Pretty funky way to integrate analytically, huh? The integration leads to a delayed (or time-shifted) step function (or constant) starting at a delayed time of $t_0 = 5$. I introduce step functions in the next section.

You can model any smooth function $x(t)$ as a series of delayed and time-shifted impulses in the following way:

$$x(t) = \int_{-\infty}^{t} x(\tau)\delta(t-\tau)\,d\tau \qquad \text{where } x(t) = x(t)u(t)$$

This equation says you can break up any function $x(t)$ into a sum of a whole bunch of delayed impulse functions with different strengths. The value of the strength is simply the function $x(t)$ evaluated where the shifted impulse occurs at time τ or t.

Stepping It Up with a Step Function

The step function is a funky function that looks like, well, a step. Practical step functions occur daily, like each time you turn mobile devices, stereos, and lights on and off. Here's the general definition of the unit step function:

$$u(t) = \begin{cases} 0 & \text{for} \quad t < 0 \\ 1 & \text{for} \quad t \geq 0 \end{cases}$$

So this step function is equal to 0 when time t is negative and is equal to 1 when time t is 0 or positive. Alternatively, you can say there's a jump in the function value at time $t = 0$. Math gurus call this jump a *discontinuity*.

Although you can't generate an ideal step function, you can approximate a step function. Figure 11-2 shows what a step function looks like, along with a circuit that's roughly a step function.

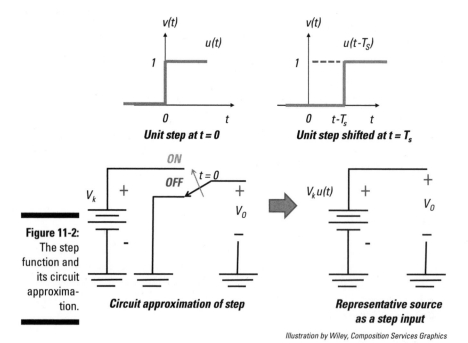

Figure 11-2: The step function and its circuit approximation.

Illustration by Wiley, Composition Services Graphics

The following sections cover some operations for shifting and weighting step functions.

Creating a time-shifted, weighted step function

The circuit approximation of the step function in Figure 11-2 assumes you can quickly change from *off* to *on* at time $t = 0$ when the switch is thrown.

Although the unit step function appears not to do much, it's a versatile signal that can build other waveforms. In a graph, you can make the step shrink or stretch. You can multiply the step function $u(t)$ by a constant amplitude V_k to produce the following waveform:

$$V_k u(t) = \begin{cases} 0 & \text{for} \quad t < 0 \\ V_k & \text{for} \quad t \geq 0 \end{cases}$$

The scale or weight of the unit input is V_k. The amplitude V_k measures the size of the jump in function value.

You can move the step function in time with a shift of T_s, leading you to a shifted, weighted waveform:

$$V_k u(t - T_s) = \begin{cases} 0 & \text{for} \quad T_s < 0 \\ V_k & \text{for} \quad T_s \geq 0 \end{cases}$$

This equation says the function equals 0 before time T_s and that the value of the function jumps to V_k after time T_s. Figure 11-3 shows the step function weighted by V_k with a time shift of T_s.

Figure 11-3:
A time-shifted step function.

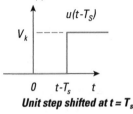

Unit step shifted at $t = T_s$

Illustration by Wiley,
Composition Services Graphics

You can add two step functions together to form a pulse function, as I show you in the next section.

Being out of step with shifted step functions

Step functions can dance around, but it's not the fancy twist-and-shout kind of dancing. The function can become bigger or smaller and move to the left or right. You can add those modified step functions to make even more funky step functions.

For example, you can generate a rectangular pulse as a sum of two step functions. To get a visual of this concept, see Figure 11-4, which shows a rectangular pulse that consists of the sum of two step functions in time. Before 1 second, the value of the pulse is 0. Then the amplitude of the pulse jumps to a value of 3 and stays at that value between 1 and 2 seconds. The pulse then returns to 0 at time $t = 2$ seconds. You wind up with the rectangular pulse $p(t)$ described as the sum of two step functions:

$$p(t) = 3u(t-1) - 3u(t-2)$$

This expression says that you create a pulse with a time-shifted step function starting at 1 second with an amplitude of 3 and add it to another time-shifted step function starting at 2 seconds with an amplitude of –3. You can view the pulse as a *gating function* for electronic switches to allow or stop a signal from passing through.

Figure 11-4: Building a rectangular pulse with step functions.

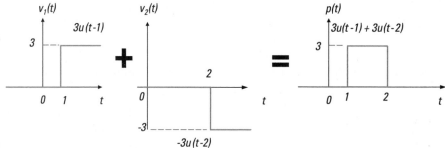

Illustration by Wiley, Composition Services Graphics

Building a ramp function with a step function

The integral of the step function generates a ramp function, which consists of two functions multiplied together:

$$r(t) = \int_{-\infty}^{t} u(x)dx$$
$$= tu(t)$$

The time function $tu(t)$ is simply a ramp function with a slope (or strength) of 1, and the unit step function serves as a convenient mathematical tool to start the ramp at time $t = 0$. You can add a strength K to the ramp and shift the ramp function in time by T_S as follows:

$$v(t) = Kr(t - T_S)$$

The ramp doesn't start until T_S. Before the time shift T_S, the ramp function is 0. After time T_S, the ramp has a value equal to $Kr(t - T_S)$.

With ramp functions, you can create triangular and sawtooth functions (or waveforms). Figure 11-5 shows a ramp of unit strength, a ramp of strength K with a time shift of 1, a triangular waveform, and a sawtooth waveform. Building such waveforms from other functions is useful when you're breaking the input into recognizable pieces and applying superposition.

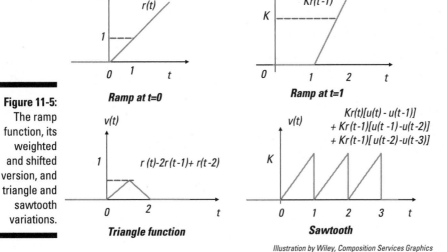

Figure 11-5: The ramp function, its weighted and shifted version, and triangle and sawtooth variations.

Illustration by Wiley, Composition Services Graphics

Here's how to build the triangle function in Figure 11-5 using ramp functions:

1. **Turn on a ramp with a slope of 1 starting at time $t = 0$.**

2. **Add a ramp that has a slope of –2 and starts at $t = 1$.**

   At $t = 1$, you see the function start to decrease with a slope of –1. But before that, the slope of the function (from the first ramp) is 1; adding a ramp with a slope of –2 to the first ramp results in a ramp with a slope of –1.

3. **Turn off the second ramp by adding another delayed ramp that has a slope of 1 and starts at time $t = 2$.**

   Adding a ramp with a slope of 1 brings the slope back to 0.

Here's the math behind what I just said:

$$v(t) = r(t) - 2r(t-1) + r(t-2)$$

Here's how to build a sawtooth function like the one in Figure 11-5 using ramp and step functions:

1. **Start with a ramp of slope (or strength) K multiplied by a rectangular pulse of unit height.**

   The pulse consists of two step functions. Mathematically, you have a ramp with a specific time duration:

   $$r_1(t) = Kr(t)[u(t) - u(t-1)]$$

2. **Apply a time delay of 1 to the ramp pulse $r_1(t)$ to get another ramp pulse $r_2(t)$ that's time shifted.**

   You get the following:

   $$r_2(t) = Kr_1(t-1) = Kr(t-1)[u(t-1) - u(t-2)]$$

3. **Repeat Step 2 to get more delayed ramp pulses starting at 2, 3, 4, and so on.**

4. **Add up all the functions to get the sawtooth $s_t(t)$.**

   Here's the sawtooth function:

   $$s_t(t) = K\{r(t)[u(t) - u(t-1)] + r(t-1)[u(t-1) - u(t-2)] + \ldots +\}$$

Pushing the Limits with the Exponential Function

The *exponential function* is a step function whose amplitude V_k gradually decreases to 0. Exponential functions are important because they're solutions to many circuit analysis problems in which a circuit contains resistors, capacitors, and inductors.

The exponential waveform is described by the following equation:

$$v(t) = V_k e^{-\left(\frac{t}{T_c}\right)} u(t)$$

The time constant T_C provides a measure of how fast the function will decay or grow. Using the step function means that the function starts at $t = 0$.

REMEMBER

A minus sign on the exponent indicates a decaying exponential, whereas a positive sign indicates a growing exponential. When you have a growing exponential, the circuit can't handle the input, and nothing works after exceeding the supplied voltage. In academia terms, the system goes *unstable*.

Here's the time-shifted version of a decaying exponential starting at time t_0:

$$v(t) = V_k e^{-\left(\frac{t-t_0}{T_C}\right)} u(t - t_0)$$

Figure 11-6 shows a decaying exponential, its time-shifted version, and a growing exponential.

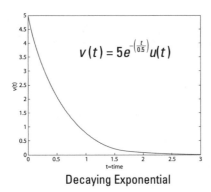

Decaying Exponential

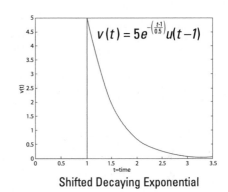

Shifted Decaying Exponential

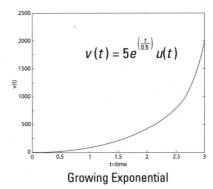

Growing Exponential

Figure 11-6:
The exponential function, its shifted version, and a growing exponential.

Illustration by Wiley, Composition Services Graphics

Seeing the Signs with Sinusoidal Functions

The sinusoidal functions (sine and cosine) appear everywhere, and they play an important role not only in electrical engineering but in many branches of science and engineering. In circuit analysis, the sinusoid serves as a good approximation to describe a circuit's input and output behavior.

The sinusoidal function is periodic, meaning its graph contains a basic shape that repeats over and over indefinitely. The function goes on forever, oscillating through endless peaks and valleys in both negative and positive directions of time. Here are some key parts of the function:

✔ The amplitude V_A defines the maximum and minimum peaks of the oscillations.

✔ Frequency f_0 describes the number of oscillations in 1 second.

✔ The period T_0 defines the time required to complete 1 cycle.

The period and frequency are reciprocals of each other, governed by the following mathematical relationship:

$$f_0 = \frac{1}{T_0}$$

In this book, I define the following cosine function as the reference signal:

$$v(t) = V_A \cos(2\pi f_0 t)$$

$$= V_A \cos\left(\frac{2\pi t}{T_0}\right)$$

You can move sinusoidal functions left or right with a time shift as well as increase or decrease the amplitude. You can also describe a sinusoidal function with a phase shift in terms of a linear combination of sine and cosine functions. Figure 11-7 shows a cosine function and a shifted cosine function with a phase shift of $\pi/2$.

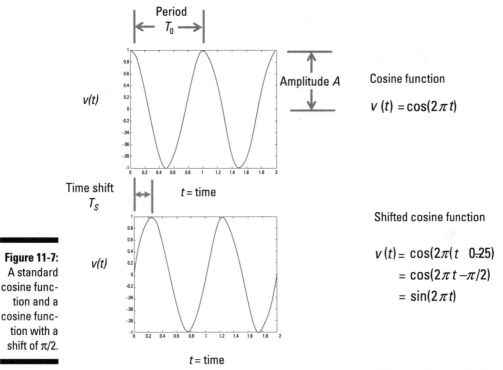

Period T_0

Amplitude A

$v(t)$

t = time

Cosine function

$$v(t) = \cos(2\pi t)$$

Time shift T_S

Figure 11-7: A standard cosine function and a cosine function with a shift of $\pi/2$.

$v(t)$

t = time

Shifted cosine function

$$v(t) = \cos(2\pi(t - 0.25)$$
$$= \cos(2\pi t - \pi/2)$$
$$= \sin(2\pi t)$$

Illustration by Wiley, Composition Services Graphics

Giving wavy functions a phase shift

A signal that's *out of phase* has been shifted left or right when compared to a reference signal:

✔ **Right shift:** When a function moves right, then the function is said to be *delayed*. The delayed cosine has its peak occur after the origin. A delayed signal is also said to be a *lag signal* because the signal arrives later than expected.

✔ **Left shift:** When the cosine function is shifted left, the shifted function is said to be *advanced*. The peak of the advanced signal occurs just before the origin. An advanced signal is also called a *lead signal* because the lead signal arrives earlier than expected.

Figure 11-8 shows unshifted, lagged, and lead cosine functions.

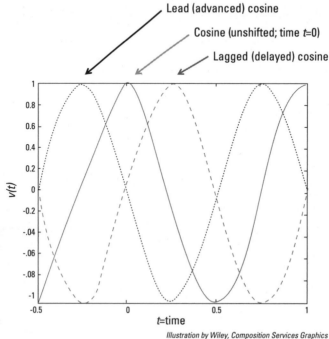

Figure 11-8:
Unshifted,
lag, and
lead cosine
functions.

To see what a phase shift looks like mathematically, first take a look at the reference signal:

$$v(t) = V_A \cos(2\pi f_0 t)$$
$$= V_A \cos\left(\frac{2\pi t}{T_0}\right)$$

At $t = 0$, the positive peak V_A serves as a reference point. To move the reference point by time shift T_S, replace the t with $(t - T_S)$:

$$v(t) = V_A \cos\left[\frac{2\pi}{T_0}(t - T_S)\right]$$
$$= V_A \cos\left[\frac{2\pi t}{T_0} - \phi\right]$$

where $\phi = 2\pi\left(\frac{T_S}{T_0}\right) = 360°\left(\frac{T_S}{T_0}\right)$.

The factor ϕ is the phase shift (or angle). The phase shift is the angle between $t = 0$ and the nearest positive peak. You can view the preceding equation as the polar representation of the sinusoid. When the phase shift is $\pi/2$, then the shifted cosine is a sine function.

Express the phase angle in radians to make sure it's in the same units as the argument of the cosine $(2\pi t/T_0 - \phi)$. *Note:* Angles can be expressed in either radians or degrees; make sure you use the right setting on your calculator.

When you have a phase shift ϕ at the output when compared to the input, it's usually caused by the circuit itself.

Expanding the function and finding Fourier coefficients

The general sinusoid $v(t)$, which I introduce in the preceding section, involves the cosine of a difference of angles. In many applications, you can expand the general sinusoid using the following trigonometric identity:

$$\cos(\alpha - \beta) = \cos\alpha\cos\beta + \sin\alpha\sin\beta$$

Expanding the general sinusoid $v(t)$ leads to

$$v(t) = V_A \cos\left[\underbrace{\frac{2\pi}{T_0}t}_{\alpha} - \underbrace{\phi}_{\beta}\right]$$

$$= \underbrace{\left[V_A\cos\phi\right]}_{c}\cos\left[\frac{2\pi t}{T_0}\right] + \underbrace{\left[V_A\sin\phi\right]}_{d}\sin\left[\frac{2\pi t}{T_0}\right]$$

The terms c and d are just special constants called *Fourier coefficients.* You can express the waveform as a combination of sines and cosines as follows:

$$v(t) = c\cos\left[\frac{2\pi t}{T_0}\right] + d\sin\left[\frac{2\pi t}{T_0}\right]$$

The function $v(t)$ describes a sinusoidal signal in rectangular form.

If you know your complex numbers going between polar and rectangular forms, then you can go between the two forms of the sinusoids. The Fourier coefficients c and d are related by the amplitude V_A and phase ϕ:

$$c = V_A \cos\phi$$
$$d = V_A \sin\phi$$

If you go back to find V_A and ϕ from the Fourier coefficients c and d, you wind up with these expressions:

$$V_A = \sqrt{c^2 + d^2}$$
$$\phi = \tan^{-1}\left(\frac{d}{c}\right)$$

The inverse tangent function on a calculator has a positive or negative $180°$ (or π) phase ambiguity. You can figure out the phase by looking at the signs of the Fourier coefficients c and d. Draw the points c and d on the rectangular system, where c is the x-component (or *abscissa*) and d is the y-component (or *ordinate*). The ratio of d/c can be negative in Quadrants II and IV. Using the rectangular system helps you determine the angles when taking the arctangent, whose range is from $-\pi/2$ to $\pi/2$.

Connecting sinusoidal functions to exponentials with Euler's formula

Euler's formula connects trig functions with complex exponential functions (see the earlier section "Pushing the Limits with the Exponential Function" for details on exponentials). The formula states that for any real number θ, you have the following complex exponential expressions:

$$e^{j\theta} = \cos\theta + j\sin\theta$$
$$e^{-j\theta} = \cos\theta - j\sin\theta$$

The exponent $j\theta$ is an imaginary number, where $j = \sqrt{-1}$. (The imaginary number j is the same as the number i from your math classes, but all the cool people use j for imaginary numbers because i stands for current.)

You can add and subtract the two preceding equations to get the following relationships:

$$\cos\theta = \frac{e^{j\theta} + e^{-j\theta}}{2}$$

$$\sin\theta = \frac{e^{j\theta} - e^{-j\theta}}{2j}$$

These equations say that the cosine and sine functions are built as a combination of complex exponentials. The complex exponentials play an important role when you're analyzing complex circuits that have storage devices such as capacitors and inductors.

Chapter 12

Spicing Up Circuit Analysis with Capacitors and Inductors

. . . .

In This Chapter

▶ Using capacitors to store electrical energy

▶ Storing magnetic energy with inductors

▶ Using op amps to do your calculus

. . . .

*I*f you've previously analyzed circuits consisting of only resistors and batteries, you may be happy to hear that more-interesting circuits do exist. The addition of two passive devices — capacitors and inductors — help spice up the functioning of circuits by storing energy for later use. You couldn't have electronic multimedia devices or entertainment gear without capacitors and inductors.

The addition of capacitors and inductors also lets you use circuits to do some calculations for you. With these devices, you can perform mathematical operations that are usually done by hand, such as integration and differentiation, electronically. Yep, you read right. You can build on-the-spot calculus operations using capacitors and resistors, along with your life-long friend, the operational amplifier (see Chapter 10 for the scoop on op amps).

In this chapter, I introduce you to capacitors and inductors, and I help you find quantities such as voltage, current, power, energy, capacitance, and inductance in circuits that contain these storage devices. I then show you how to do a little calculus with op amps.

Storing Electrical Energy with Capacitors

Interesting things happens when capacitors come into play in circuit analysis. They allow you to build voltage dividers that depend on the frequency content of the signals. What use is that? Well, with capacitors and resistors, you can emphasize the frequencies produced by specific instruments in your favorite music, like the high-frequency beats from a snare drum or the low-frequency bass sounds of a cello. Or you can filter out the voices in a song to create your own karaoke soundtrack.

Other uses for capacitors include filtering and storing energy by bypassing or coupling capacitors to make circuits work properly.

The following sections give you insight into capacitors and the relationship between voltage and current in a capacitor. They also explain how to find the amount of energy stored in a capacitor, whether you're dealing with a single capacitor or multiple capacitors in a parallel or series construction.

Describing a capacitor

A capacitor consists of two parallel conducting plates like silver or aluminum separated by an insulator. Unlike resistors, which waste energy, capacitors store energy for later use. Here's the property that applies to capacitors: $q = Cv$. C is the capacitance, q is the amount of stored charge, and v is the voltage across the capacitor.

The *capacitance,* which is measured in *farads* (F), relates the amount of charge stored in a capacitor to the applied voltage. The formula shows that the larger the voltage across a capacitor, the larger the amount of stored charges. How much larger depends on the capacitance value. Because voltage is the amount of energy per unit charge, capacitance also measures a capacitor's ability to store energy. The larger the capacitance, the more energy a capacitor can store. You can vary the amount of charge stored in a capacitor by changing certain physical properties of a capacitor, like the area of the conducting plates or their distance apart.

Figure 12-1 shows the schematic symbol for a capacitor: two parallel lines of equal length, separated by a gap. If you see a plus sign by the symbol, the capacitor is polarized. Polarized capacitors show distinct polarities; they're touchy in how you should connect the voltage polarities to the circuit.

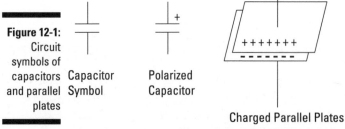

Figure 12-1:
Circuit symbols of capacitors and parallel plates

Capacitor Symbol

Polarized Capacitor

Charged Parallel Plates

Illustration by Wiley, Composition Services Graphics

Charging a capacitor (credit cards not accepted)

When you connect a battery to a capacitor, the negative side of the battery pushes negative charges on one of the plates. These electrons form an electric field, repelling electrons on the other plate and leaving a positive charge. The electrons build up according to the amount of applied voltage. If the applied voltage remains constant, then the electrons build up until there's no current flow. You now have a *charged* capacitor.

If you disconnect the charged capacitor from the battery, the capacitor saves the same voltage. Even more magic occurs when you connect a charged capacitor to a circuit with resistors. The voltage across the capacitor releases charges (or current) to discharge the capacitor. Eventually, the capacitor discharges to 0 volts for a circuit with no voltage sources. This charging and discharging action occurs over time, and because the action takes time, you can use capacitors in timing applications, like triggering an alarm to remind you to take a break to do 100 push-ups.

Relating the current and voltage of a capacitor

The voltage and current of a capacitor are related. To see this, you need to take the derivative of the capacitance equation $q(t) = Cv(t)$, which is

$$\frac{dq(t)}{dt} = C\frac{dv(t)}{dt}$$

Because $dq(t)/dt$ is the current through the capacitor, you get the following *i-v* relationship:

$$i(t) = C\frac{dv(t)}{dt}$$

This equation tells you that when the voltage doesn't change across the capacitor, current doesn't flow; to have current flow, the voltage must change. For a constant battery source, capacitors act as open circuits because there's no current flow.

The voltage across a capacitor changes in a smooth fashion (and its derivatives are also smoothly changing functions), so there are no instantaneous jumps in voltages.

Just as you don't have gaps in velocities when you accelerate or deceler-
ate your car, you don't have gaps in voltages. The mass of the car causes a
smooth transition when going from 55 miles per hour to 60 miles per hour. In
a similar and analogous way, you can think of the capacitance C as the mass in
the circuit world that causes a smooth transition when changing voltages from
one value to another.

To express the voltage across the capacitor in terms of the current, you integrate the preceding equation as follows:

$$v(t) = \frac{1}{C}\int_0^t i(\tau)d\tau + v_c(0)$$

The second term in this equation is the initial voltage across the capacitor at
time $t = 0$.

You can see the i-v characteristic in Figure 12-2. The left diagram defines a
linear relationship between the charge q stored in the capacitor and the voltage v across the capacitor. The right diagram shows a current relationship
between the current and the derivative of the voltage, $dv_c(t)/dt$, across the
capacitor with respect to time t.

Think of capacitance C as a proportionality constant, like a resistor acts as a
constant in Ohm's law.

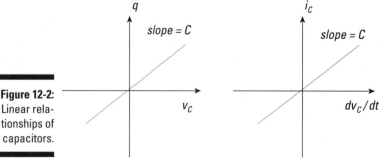

Figure 12-2:
Linear rela-
tionships of
capacitors.

Illustration by Wiley, Composition Services Graphics

Finding the power and energy of a capacitor

To find the instantaneous power of the capacitor, you need the following
power definition, which applies to any device:

$$p_C(t) = \underbrace{i_C(t)}_{C\frac{dv_C(t)}{dt}} v_C(t)$$

The subscript C denotes a capacitance device (surprise!). Substituting the current for a capacitor (from the preceding section) into this equation gives you the following:

$$p_C(t) = C\frac{dv_C(t)}{dt}v_C(t)$$

$$\frac{dw_C(t)}{dt} = \frac{d}{dt}\left[\frac{1}{2}Cv_C^2(t)\right]$$

Assuming zero initial voltage, the energy $w_C(t)$ stored per unit time is the power. Integrating that equation gives you the energy stored in a capacitor:

$$w_C(t) = \frac{1}{2}Cv_C^2(t)$$

The energy equation implies that the energy stored in a capacitor is always positive. The capacitor absorbs power from a circuit when storing energy. The capacitor releases the stored energy when delivering energy to the circuit.

For a numerical example, look at the top-left diagram of Figure 12-3, which shows how the voltage changes across a 0.5-µF capacitor. Try calculating the capacitor's energy and power.

The slope of the voltage change (time derivative) is the amount of current flowing through the capacitor. Because the slope is constant, the current through the capacitor is constant for the given slopes. For this example, you calculate the slope for each time interval in the graph as follows:

$$\frac{dv_C(t)}{dt} = \frac{(10-0)\text{ V}}{(0.002-0)\text{ s}} = 5{,}000\text{ V/s} \qquad 0 \le t < 2\text{ ms}$$

$$\frac{dv_C(t)}{dt} = \frac{(10-10)\text{ V}}{(0.004-0.002)\text{ s}} = 0\text{ V/s} \qquad 2\text{ ms} \le t < 4\text{ ms}$$

$$\frac{dv_C(t)}{dt} = \frac{(0-10)\text{ V}}{(0.006-0.004)\text{ s}} = -5{,}000\text{ V/s} \qquad 4\text{ ms} \le t$$

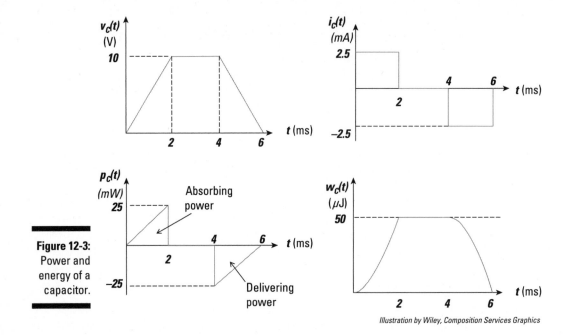

Figure 12-3:
Power and
energy of a
capacitor.

Illustration by Wiley, Composition Services Graphics

Multiply the slopes by the capacitance (in farads) to get the capacitor current during each interval. The capacitance is 0.5 μF, or 0.5×10^{-6} F, so here are the currents:

$$i_C(t) = C\frac{dv(t)}{dt}$$

$$i_C(t) = \left(0.5 \times 10^{-6} \text{ F}\right)(5{,}000 \text{ V/s}) = 2.5 \text{ mA} \qquad 0 \leq t < 2 \text{ ms}$$

$$i_C(t) = \left(0.5 \times 10^{-6} \text{ F}\right)(0 \text{ V/s}) = 0 \text{ mA} \qquad 2 \text{ ms} \leq t < 4 \text{ ms}$$

$$i_C(t) = \left(0.5 \times 10^{-6} \text{ F}\right)(-5{,}000 \text{ V/s}) = -2.5 \text{ mA} \qquad 4 \text{ ms} \leq t$$

You see the graph of the calculated currents in the top-right diagram of Figure 12-3.

You find the power by multiplying the current and voltage, resulting in the bottom-left graph in Figure 12-3. Finally, you can find the energy by calculating $(1/2)C[v_C(t)]^2$. When you do this, you get the bottom-right graph of Figure 12-3. Here, the capacitor's energy increases when it's absorbing power and decreases when it's delivering power.

Calculating the total capacitance for parallel and series capacitors

You can reduce capacitors connected in parallel or connected in series to one single capacitor. This section shows you how.

Finding the equivalent capacitance of parallel capacitors

Consider the first circuit in Figure 12-4, which contains three parallel capacitors. Because the capacitors are connected in parallel, they have the same voltages:

$$v_1(t) = v_2(t) = v_3(t) = v(t)$$

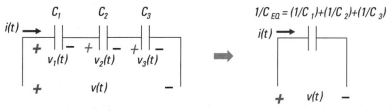

Figure 12-4: Parallel and series connection of capacitors.

Illustration by Wiley, Composition Services Graphics

Adding the current from each parallel capacitor gives you the net current $i(t)$:

$$i(t) = \underbrace{C_1 \frac{dv(t)}{dt}}_{i_1(t)} + \underbrace{C_2 \frac{dv(t)}{dt}}_{i_2(t)} + \underbrace{C_3 \frac{dv(t)}{dt}}_{i_3(t)}$$

$$= \underbrace{(C_1 + C_2 + C_3)}_{C_{EQ}} \frac{dv(t)}{dt}$$

For parallel capacitors, the equivalent capacitance is

$$C_{EQ} = C_1 + C_2 + C_3$$

Finding the equivalent capacitance of capacitors in series

For a series connection of capacitors, apply Kirchhoff's voltage law (KVL) around a loop in the bottom diagram of Figure 12-4. KVL says the sum of the voltage rises and drops around a loop is 0, giving you

$$v(t) = \underbrace{v_1(t)}_{v_1(0)+\frac{1}{C_1}\int_0^t i_1(\tau)d\tau} + \underbrace{v_2(t)}_{v_2(0)+\frac{1}{C_2}\int_0^t i_2(\tau)d\tau} + \underbrace{v_3(t)}_{v_3(0)+\frac{1}{C_3}\int_0^t i_3(\tau)d\tau}$$

A series current has the same current $i(t)$ going through each of the series capacitors, so

$$v(t) = \left[v_1(0) + v_2(0) + v_3(0)\right] + \underbrace{\left(\frac{1}{C_1} + \frac{1}{C_2} + \frac{1}{C_3}\right)}_{\frac{1}{C_{EQ}}}\int_0^t i(\tau)d\tau$$

The preceding equation shows how you can reduce the series capacitance to one single capacitance:

$$\frac{1}{C_{EQ}} = \frac{1}{C_1} + \frac{1}{C_2} + \frac{1}{C_3}$$

Storing Magnetic Energy with Inductors

Inductors find heavy use in radiofrequency (RF) circuits. They serve as RF "chokes," blocking high-frequency signals. This application of inductor circuits is called *filtering*. Electronic filters select or block whichever frequencies the user chooses.

In the following sections, you discover how inductors resist instantaneous changes in current and store magnetic energy. You also find the equivalent inductance when inductors are connected in series or in parallel.

Describing an inductor

Unlike capacitors, which are electrostatic devices, inductors are electromagnetic devices. Whereas capacitors avoid an instantaneous change in voltage, inductors prevent an abrupt change in current. Inductors are wires wound into several loops to form coils. In fact, the inductor's symbol looks like a coil of wire (see Figure 12-5).

Figure 12-5:
The circuit symbol for an inductor.

Circuit
Symbol

Illustration by Wiley, Composition Services Graphics

Current flowing through a wire creates a magnetic field, and the magnetic field lines encircle the wire along its axis. The concentration, or density, of the magnetic field lines is called *magnetic flux*. The coiled shape of inductors increases the magnetic flux that naturally occurs when current flows through a straight wire. The greater the flux, the greater the inductance. You can get even larger inductance values by inserting iron into the wire coil.

Here's the defining equation for the inductor:

$$v_L(t) = L\frac{di_L(t)}{dt}$$

where the inductance L is a constant measured in *henries* (H). You see this equation in graphical form in Figure 12-6. The figure shows the *i-v* characteristic of an inductor, where the slope of the line is the value of the inductance.

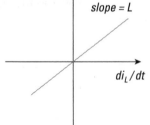

Figure 12-6:
Linear relationship of inductors.

Illustration by Wiley, Composition Services Graphics

The preceding equation says that the voltage across the inductor depends on the time rate of change of the current. In other words, no change in inductor current means no voltage across the inductor. To create voltage across the inductor, current must change smoothly. Otherwise, an instantaneous change in current would create one humongous voltage across the inductor.

Think of inductance L as a proportionality constant, like a resistor acts as a constant in Ohm's law. This notion of Ohm's law for inductors (and capacitors) becomes useful when you start working with phasors (see Chapter 15).

To express the current through the inductor in terms of the voltage, you integrate the preceding equation as follows:

$$i_L(t) = \frac{1}{L}\int_0^t v(\tau)d\tau + i_L(0)$$

The second term in this equation is the initial current through the inductor at time $t = 0$.

Finding the energy storage of an attractive inductor

To find the energy stored in the inductor, you need the following power definition, which applies to any device:

$$p_L(t) = \underbrace{v_L(t)}_{L\frac{di_L(t)}{dt}}i_L(t)$$

The subscript L denotes an inductor device. Substituting the voltage for an inductor (from the preceding section) into the power equation gives you the following:

$$p_L(t) = L\frac{di_L(t)}{dt}i_L(t)$$

$$\frac{dw_L(t)}{dt} = \frac{d}{dt}\left[\frac{1}{2}Li_L^2(t)\right]$$

The energy $w_L(t)$ stored per unit time is the power. Integrating the preceding equation gives you the energy stored in an inductor:

$$w_L(t) = \frac{1}{2}Li_L^2(t)$$

The energy equation implies that the energy in the inductor is always positive. The inductor absorbs power from a circuit when storing energy, and the inductor releases the stored energy when delivering energy to the circuit.

To visualize the current and energy relationship, consider Figure 12-7, which shows the current as a function of time and the energy stored in an inductor. The figure also shows how you can get the current from the inductor relationship between current and voltage.

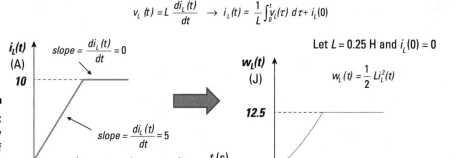

Figure 12-7:
Energy
storage of
inductors.

Illustration by Wiley, Composition Services Graphics

Calculating total inductance for series and parallel inductors

Inductors connected in series or connected in parallel can be reduced to one single inductor, as I explain next.

Finding the equivalent inductance for inductors in series

Take a look at the circuit with three series inductors shown in the top diagram of Figure 12-8. Because the inductors are connected in series, they have the same currents:

$$i_1(t) = i_2(t) = i_3(t) = i(t)$$

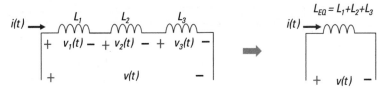

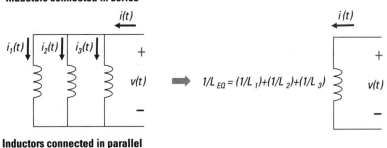

Figure 12-8: Total inductance for inductors connected in series and parallel.

Illustration by Wiley, Composition Services Graphics

Add up the voltages from the series inductors to get the net voltage $v(t)$, as follows:

$$v(t) = L_1 \underbrace{\frac{di(t)}{dt}}_{v_1(t)} + L_2 \underbrace{\frac{di(t)}{dt}}_{v_2(t)} + L_3 \underbrace{\frac{di(t)}{dt}}_{v_3(t)}$$

$$= \underbrace{(L_1 + L_2 + L_3)}_{L_{EQ}} \frac{di(t)}{dt}$$

For a series inductors, you have an equivalent inductance of

$$L_{EQ} = L_1 + L_2 + L_3$$

Finding the equivalent inductance for inductors in parallel

For a parallel connection of inductors, apply Kirchhoff's current law (KCL) in the bottom diagram of Figure 12-8. KCL says the sum of the incoming currents and outgoing current at a node is equal to 0, giving you

$$i(t) = \underbrace{i_1(t)}_{i_1(0)+\frac{1}{L_1}\int_0^t v_1(\tau)d\tau} + \underbrace{i_2(t)}_{i_2(0)+\frac{1}{L_2}\int_0^t v_2(\tau)d\tau} + \underbrace{i_3(t)}_{i_3(0)+\frac{1}{L_3}\int_0^t v_3(\tau)d\tau}$$

Because you have the same voltage $v(t)$ across each of the parallel inductors, you can rewrite the equation as

$$i(t) = \left[i_1(0) + i_2(0) + i_3(0)\right] + \underbrace{\left(\frac{1}{L_1} + \frac{1}{L_2} + \frac{1}{L_3}\right)}_{\frac{1}{L_{EQ}}}\int_0^t v(\tau)d\tau$$

This equation shows how you can reduce the parallel inductors to one single inductor:

$$\frac{1}{L_{EQ}} = \frac{1}{L_1} + \frac{1}{L_2} + \frac{1}{L_3}$$

Calculus: Putting a Cap on Op-Amp Circuits

In this section, you add a capacitor to an operational-amplifier (op-amp) circuit. Doing so lets you use the circuit to do more-complex mathematical operations, like integration and differentiation. Practically speaking, you use capacitors instead of inductors because inductors are usually bulkier than capacitors.

Creating an op-amp integrator

Figure 12-9 shows an op-amp circuit that has a feedback element as a capacitor. The circuit is configured similarly to an inverting amplifier. (Check out Chapter 10 if you want to brush up on op amps.)

The cool thing about this op-amp circuit is that it performs integration. The circuit electronically calculates the integral of any input voltage, which is a lot simpler (and less painful!) than banging your head on the table as you try to integrate a weird function by hand.

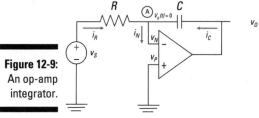

Figure 12-9:
An op-amp
integrator.

Illustration by Wiley, Composition Services Graphics

I walk you through the analysis so you can see how this circuit performs this incredible feat called integration. First, you use a KCL equation at Node A:

$$i_R(t) + i_C(t) = i_N(t)$$

Ohm's law ($i = v/R$) gives you the current through the resistor:

$$i_R(t) = \frac{1}{R}\left[v_S(t) - v_G(t)\right]$$

You get the current through the capacitor using the *i-v* relationship of a capacitor:

$$i_C(t) = C\frac{d\left[v_O(t) - v_G(t)\right]}{dt}$$

For ideal op-amp devices (see Chapter 10), the circuit gives you $v_G(t) = 0$ (virtual ground) and $i_N = 0$ (infinite input resistance). Substituting these op-amp constraints for $i_R(t)$ and $i_C(t)$ into the KCL equation gives you

$$\frac{v_S(t)}{R} + C\frac{dv_O(t)}{dt} = 0$$

Then integrate both sides of the preceding equation. You wind up with the following output voltage $v_o(t)$:

$$v_O(t) = -\frac{1}{RC}\int_0^t v_s(\tau)d\tau + v_o(0)$$

The initial output voltage $v_o(0)$ across the capacitor — that, is the voltage at $t = 0$ — is 0. If $v_o(0) = 0$, then the output-voltage equation reduces to

$$v_O(t) = -\frac{1}{RC}\int_0^t v_S(\tau)d\tau$$

The op-amp circuit accepts an input voltage and gives you an inverted output that's proportional to the integral of the input voltage.

Deriving an op-amp differentiator

With op-amp circuits where the resistor is the feedback element and the capacitor is the input device (like the one in Figure 12-10), you can perform differentiation electronically.

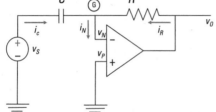

Figure 12-10:
An op-amp
differentiator.

Illustration by Wiley, Composition Services Graphics

You follow the same process as the one you use to find the relationship for an op-amp integrator (see the preceding section for details). Begin with a KCL equation at Node G:

$$i_R(t) + i_C(t) = i_N(t)$$

The current through the resistor is given by Ohm's law ($i = v/R$):

$$i_R(t) = \frac{1}{R}\left[v_O(t) - v_G(t)\right]$$

The current through the capacitors is given by the *i-v* relationship of a capacitor:

$$i_C(t) = C \frac{d[v_S(t) - v_G(t)]}{dt}$$

For ideal op-amp devices (see Chapter 11), the circuit gives you $v_G = 0$ (virtual ground) and $i_N = 0$ (infinite input resistance). Substituting these op-amp constraints for $i_R(t)$ and $i_C(t)$ into the KCL equation gives you the following:

$$\frac{v_0(t)}{R} + C \frac{dv_S(t)}{dt} = 0$$

Solving for v_o, you wind up with the following output voltage $v_o(t)$:

$$v_0(t) = -RC \frac{dv_S(t)}{dt}$$

So if you give me an input voltage, I say no sweat in getting its derivative as an output. The inverted output is simply proportional to the derivative of the input voltage.

Using Op Amps to Solve Differential Equations Really Fast

The op-amp circuit can solve mathematical equations fast, including calculus problems. The intent of this section is to give you a basic idea of how to implement various op-amp configurations and how they can be tied together.

Say you want to solve a differential equation by finding $v(t)$, a function that's a solution to a differential equation. In the following example, I show you how to use various op-amp configurations to find the output voltage $v_o(t) = v(t)$.

To simplify the problem, assume zero initial conditions: zero initial capacitor voltage for each integrator in Figure 12-11. To solve a differential equation, you need to develop a block diagram for the differential equation (which is represented by the dashed boxes in the figure), giving the input and the output for each dashed box. Then use the block diagram to design a circuit.

On the far left of Figure 12-11 is a forcing function of 25 volts derived from the following steps, and the output voltage $v_o(t) = v(t)$ is on the far right of the figure.

$$\frac{d^2v}{dt^2} = -20\frac{dv}{dt} - 100\ v + 25\ V$$

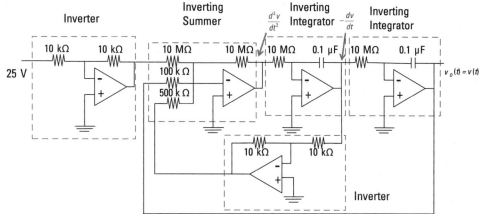

Figure 12-11:
Solving differential equations with op amps.

Here are the basic steps for designing the circuit:

1. **Solve for the highest-order derivative, showing that it consists of a sum of the lower derivatives.**

   Suppose you want to solve the following second-order differential equation:

   $$10\frac{d^2v(t)}{dt^2} + 200\frac{dv(t)}{dt} + 1000v(t) = 250\ V$$

   The first step is to algebraically solve for the highest-order derivative, d^2v/dt^2:

   $$\frac{d^2v}{dt^2} = -20\frac{dv}{dt} - 100v + 25\ V$$

The highest-order derivative is a combination or sum of lower derivatives and the smaller input voltage: dv/dt, v, and 25. Therefore, you need an inverting summer to add the three terms, and these terms are forcing functions (or inputs) to the inverting summer.

2. **Use integrators to help implement the block diagram, because the integral of the higher-order derivative is the derivative that's one order lower.**

For this example, integrate the second derivative, d^2v/dt^2, to give you the first derivative, dv/dt. As Figure 12-11 shows, the output of the inverting summing amplifier is the second derivative (which is also the input to the first integrator). The output of the first inverting integrator is the negative of the first derivative dv/dt and serves as the input to the second inverting integrator. With the second inverting integrator in Figure 12-11, integrate the negative of the first derivative, $-dv/dt$, to give you the desired output, $v(t)$.

3. **Take the outputs of the integrators, scale them, and feed them back to a summer (summing amplifier).**

The second derivative consists of a sum of three terms, so this is where the op-amp inverting summer comes in:

   1. One of the inputs is a constant of 25 volts to the summer and will be an input voltage (or driving) source. The 25 volts at the input is fed to one of the inputs to the summer with a gain of 1.

   2. The output of the first integrator is the first derivative of $v(t)$, which has a weight of 20 and is fed to the second input of the inverting summer.

   3. The output of the second integrator is fed to the third input to the inverting summer with a weight of 100.

This completes the block diagram.

For this example, multiply the first derivative dv/dt by -10 and multiply v by -100. Sum them as shown in the block diagram of Figure 12-11.

4. **Design the circuit to implement the block diagram.**

To simplify the design, give each integrator a gain of -1. You need two more inverting amplifiers to make the signs come out right. Use the summer to achieve the gains of -10 and -100 found in Step 3. Figure 12-11 is one of many possible designs.

Chapter 13

Tackling First-Order Circuits

In This Chapter

▶ Focusing on first-order differential equations with constant coefficients

▶ Analyzing a series circuit that has a single resistor and capacitor

▶ Analyzing a parallel circuit that has a single resistor and inductor

*B*uilding more exciting circuit functions requires capacitors or inductors. These storage devices — which I introduce in Chapter 12 — tell other parts of the circuit to slow down and take time when things are about to change. Nothing happens instantaneously with capacitors and inductors. You can think of these devices as little bureaucracies slowing things down in the life of circuit city.

This chapter focuses on circuits with a single storage element connected to a single resistor or a resistor network. In math mumbo jumbo, a circuit with a single storage device is described with first-order differential equations; hence the name *first-order circuit*. The analysis helps you understand timing circuits and time delays if that's what's needed to achieve specific tasks. (A circuit with two storage devices is a second-order circuit, which I cover in Chapter 14.)

If your head is cloudy on the calculus, check out a diff EQ textbook or pick up a copy of *Differential Equations For Dummies* by Steven Holzner (Wiley) for a refresher.

Solving First-Order Circuits with Diff EQ

To find out what's happening in circuits with capacitors, inductors, and resistors, you need differential equations. Why? Because generating current through a capacitor requires a change in voltage, and generating voltage across an inductor requires a change in current. Differential equations take the rate of change into account.

You have a *first-order circuit* when a first-order differential equation describes the circuit. A resistor and capacitor connected in series (an *RC series circuit*) is one example of a first-order circuit. A capacitor's version of Ohm's law with capacitance C is described by a first-order derivative:

$$i_C(t) = C\frac{dv_C(t)}{dt}$$

Another first-order circuit is a resistor and inductor connected in parallel (an *RL parallel circuit*). An inductor with inductance value L has an *i-v* relationship also expressed by a first-order derivative:

$$v_L(t) = L\frac{di_L(t)}{dt}$$

Because the capacitance C and inductance L are constant and connected to a constant resistor, circuits with these devices lead to differential equations that have constant coefficients.

So when analyzing a circuit with an inductor or a capacitor with a resistor driven by an input source, you have a first-order differential equation. Both types of first-order circuits have only one energy storage device and one resistor, which converts electricity to heat. To get a complete solution to the first-order differential equation, you need to know a circuit's initial condition. An *initial condition* is simply the initial state of the circuit, such as the inductor current or the capacitor voltage at time $t = 0$.

You can solve differential equations in numerous ways, but because the circuits you encounter in this book have only constant values of resistors, inductors, and capacitors — leading to differential equations with constant coefficients — I give you just one approach to solving first-order differential. (This approach works for solving second-order differential equations, too, but that's the subject of Chapter 14.) The best part about this approach is that it converts a problem involving a differential equation to one that only involves algebra.

Here's how to solve a differential equation that has constant coefficients for first-order circuits, given an initial condition and *forcing function* (an input signal or function):

1. **Find the zero-input response by setting the input source to 0.**

   You want the output to be due to initial conditions only.

2. **Find the zero-state response by setting the initial conditions equal to 0 and adding together the solutions to the homogeneous equation and differential equation to a particular input.**

   You want the output to be due to the input signal, or forcing function, only. In first-order circuits, you have 0 initial capacitor voltage or 0 initial inductor current.

   To get the *zero-state response,* you have to find the following:

   - **The homogeneous solution:** You get the solution to the homogeneous differential equation when you first set the input signal or forcing function equal to 0. This solution is for the zero initial condition.

   - **The particular solution:** The particular solution is the solution to the differential equation with a particular input source. This means turning the input signal back on, so the solution depends on the type of input signal. For example, if your input is a constant, then your particular solution is also a constant. When you have a sine or cosine function as an input, the output is a combination of sine and cosine functions.

3. **Add up the zero-input and zero-state responses to get the total response.**

   Because you're dealing with linear circuits, you can add up the two solutions based on the superposition concept, which I cover in Chapter 7.

The following sections show you how to find the solution to a first-order differential equation. You start off with one circuit having a zero-input source and then look at circuits with a particular input source like a step input.

Guessing at the solution with the natural exponential function

Say you want to solve a homogeneous differential equation with constant coefficients having a zero-input source. The solution results from only the initial state (or initial condition) of the circuit. This response is called the *zero-input response.*

Consider the following homogeneous differential equation with zero forcing function $v_s(t) = 0$:

$$10\frac{dv(t)}{dt} + 20v(t) = v_s(t) = 0$$

The function $v_h(t)$ is the solution to the homogeneous differential equation.

You need to guess the function $v(t)$ to get 0 for the differential equation. Pssst . . . try $v(t) = e^{kt}$. Why? Because each time you take its derivative, you get the same exponential multiplied by constant k. When you substitute $v(t)$ into the differential equation, adding up the combination of exponentials leads to 0. The exponential equation is your best friend when solving this type of differential equation with constant coefficients.

The integral of an exponential is also an exponential multiplied by some constant or scale factor. This property makes exponentials useful in circuit analysis and many other applications.

Using the characteristic equation for a first-order equation

You can convert a first-order differential equation to a problem that involves algebra. Here's how you do it. You start with the zero forcing function (which I introduce in the preceding section):

$$10\frac{dv(t)}{dt} + 20v(t) = 0$$

Substitute your best-guess solution, $v(t) = e^{kt}$ (also from the preceding section), into the differential equation. With a little factoring, you get

$$10\frac{d}{dt}\left(e^{kt}\right) + 20e^{kt} = 0$$
$$(10k + 20)e^{kt} = 0$$

The coefficient of e^{kt} must be 0. Use that idea to find k:

$$10k + 20 = 0$$
$$k = -\frac{1}{2}$$

Setting the algebraic equation to 0 gives you a *characteristic equation*. Why? Because solving for the root k determines the features of the solution $v(t)$. Here, the characteristic root found as $k = -1/2$ gives you the homogeneous solution:

$$v(t) = Ae^{-\left(\frac{1}{2}\right)t}$$

You determine the constant A by applying the initial condition or state $v(0)$ when $t = 0$. Guessing at a reasonable solution to the differential equation leads to a simpler, algebraic characteristic equation.

Analyzing a Series Circuit with a Single Resistor and Capacitor

A first-order *RC series circuit* has one resistor (or network of resistors) and one capacitor connected in series. You can see an example of one in Figure 13-1. In the following sections, I show you how to find the total response for this circuit.

If your RC series circuit has a capacitor connected with a network of resistors rather than a single resistor, you can use the same approach to analyze the circuit. You just have to find the Thévenin equivalent first, reducing the resistor network to a single resistor in series with a single voltage source. See Chapter 8 for details on the Thévenin approach.

Figure 13-1:
A first-order
RC series
circuit.

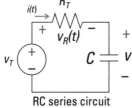

RC series circuit

Illustration by Wiley, Composition Services Graphics

Starting with the simple RC series circuit

The simple RC series circuit in Figure 13-1 is driven by a voltage source. Because the resistor and capacitor are connected in series, they must have the same current $i(t)$. For Figure 13-1 and what follows next, let $R=R_T$.

To find the voltage across the resistor $v_R(t)$, you use Ohm's law for a resistor device:

$$v_R(t) = Ri(t)$$

The element constraint for a capacitor (found in Chapter 12) is given as

$$i(t) = C\frac{dv(t)}{dt}$$

where $v(t)$ is the capacitor voltage.

Generating current through a capacitor takes a changing voltage. If the capacitor voltage doesn't change, the current in the capacitor equals 0. Zero current implies infinite resistance for constant voltage across the capacitor.

Now substitute the capacitor current $i(t) = Cdv(t)/dt$ into Ohm's law for resistor R, because the same current flows through the resistor and capacitor. This gives you the voltage across the resistor, $v_R(t)$:

$$v_R(t) = RC\frac{dv(t)}{dt}$$

Kirchhoff's voltage law (KVL) says the sum of the voltage rises and drops around a loop of a circuit is equal to 0. Using KVL for the RC series circuit in Figure 13-1 gives you

$$v_T(t) = v_R(t) + v(t)$$

Now substitute $v_R(t)$ into KVL:

$$v_T(t) = RC\frac{dv(t)}{dt} + v(t)$$

You now have a first-order differential equation where the unknown function is the capacitor voltage. Knowing the voltage across the capacitor gives you the electrical energy stored in a capacitor.

In general, the capacitor voltage is referred to as a *state variable* because the capacitor voltage describes the state or behavior of the circuit at any time.

An easy way to remember that state variables — such as the capacitor voltage $v_C(t)$ and inductor current $i_L(t)$ — describe the present situation of the circuit is to think of your car's position and instantaneous velocity as your car's state variables. If you're racing along the majestic road of Rocky Mountain National Park, your GPS position and car's speed describe the current state of your driving.

The RC series circuit is a first-order circuit because it's described by a first-order differential equation. A circuit reduced to having a single equivalent capacitance and a single equivalent resistance is also a first-order circuit. The circuit has an applied input voltage $v_T(t)$.

To find the total response of an RC series circuit, you need to find the zero-input response and the zero-state response and then add them together. Figure 13-2 shows an RC series circuit broken up into two circuits. The top-right diagram shows the zero-input response, which you get by setting the input to 0. The bottom-right diagram shows the zero-state response, which you get by setting the initial conditions to 0.

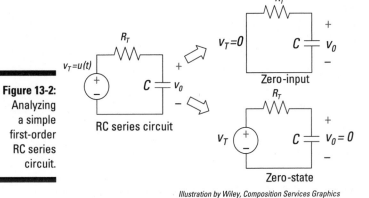

Figure 13-2: Analyzing a simple first-order RC series circuit.

Illustration by Wiley, Composition Services Graphics

Finding the zero-input response

You first want to find the zero-input response for the RC series circuit. The top-right diagram of Figure 13-2 shows the input signal $v_T(t)$ equal to 0. Zero-input voltage means you have zero . . . nada . . . zip . . . input for all time. The output response is due to the initial condition V_0 (initial capacitor voltage) at time $t = 0$. The first-order differential equation reduces to

$$v_T(t) = 0 = RC\frac{dv_{ZI}(t)}{dt} + v_{ZI}(t) \quad \text{or} \quad v_{ZI}(t) = -RC\frac{dv_{ZI}(t)}{dt}$$

Here, $v_{ZI}(t)$ is the capacitor voltage. For an input source set to 0 volts in Figure 13-2, the capacitor voltage is called a *zero-input response* or *free response*. No external forces (such as a battery) are acting on the circuit, except for the initial state of the capacitor voltage.

You can reasonably guess that the solution is the exponential function (you can check and verify the solution afterward). You try an exponential because the time derivative of an exponential is also an exponential (as I explain earlier in "Guessing at the solution with the natural exponential function"). Substitute that guess into the RC first-order circuit equation:

$$v_{ZI}(t) = Ae^{kt}$$

The A and k are arbitrary constants of the zero-input response.

Now substitute the solution $v_{ZI}(t) = Ae^{kt}$ into the differential equation:

$$v_{ZI}(t) = -RC\frac{dv_{ZI}(t)}{dt}$$

$$Ae^{kt} = -RC\frac{d(Ae^{kt})}{dt}$$

$$Ae^{kt} = -RC(kAe^{kt})$$

You get an algebraic characteristic equation after setting the equation equal to 0 and factoring out Ae^{kt}:

$$Ae^{kt}(1 + RCk) = 0$$

The characteristic equation gives you a much simpler problem. The coefficient of e^{kt} has to be 0, so you just solve for the constant k:

$$1 + RCk = 0$$

$$k = -\frac{1}{RC}$$

When you have k, you have the zero-input response $v_{ZI}(t)$. Using $k = -1/RC$, you can find the solution to the differential equation for the zero input:

$$v_{ZI}(t) = Ae^{-\left(\frac{t}{RC}\right)}$$

Now you can find the constant A by applying the initial condition. At time $t = 0$, the initial voltage is V_0, which gives you

$$v_{ZI}(0) = Ae^{-\left(\frac{0}{RC}\right)}$$

$$= A = V_0$$

The constant A is simply the initial voltage V_0 across the capacitor.

Finally, you have the solution to the capacitor voltage, which is the zero-input response $v_{ZI}(t)$:

$$v_{ZI}(t) = V_0 e^{-\left(\frac{t}{RC}\right)} = v(0) e^{-\left(\frac{t}{RC}\right)} \quad \text{where} \quad V_0 = v(0)$$

The constant term RC in this equation is called the *time constant*. The time constant provides a measure of how long a capacitor has discharged or charged. In this example, the capacitor starts at some initial state of voltage V_0 and dissipates quietly into oblivion to another state of 0 volts.

Suppose RC = 1 second and initial voltage V_0 = 5 volts. Figure 13-3 plots the decaying exponential, showing that it takes about 5 time constants, or 5 seconds, for the capacitor voltage to decay to 0.

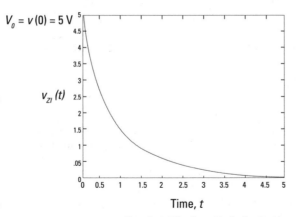

Figure 13-3:
Zero-input
response
and the
natural
exponential.

Illustration by Wiley, Composition Services Graphics

Finding the zero-state response by focusing on the input source

Zero-state response means zero initial conditions, and it requires finding the capacitor voltage when there's an input source, $v_T(t)$. You need to find the homogeneous and particular solutions to get the zero-state response. To find zero initial conditions, you look at the circuit when there's no voltage across the capacitor at time t = 0.

The circuit at the bottom right of Figure 13-2 has *zero initial conditions* and an input voltage of $V_T(t) = u(t)$, where $u(t)$ is a unit step input. Mathematically, you can describe step function $u(t)$ as

$$u(t) = \begin{cases} 0 & t < 0 \\ 1 & t \geq 0 \end{cases}$$

The input signal is divided into two time intervals. When $t < 0$, $u(t) = 0$. The first-order differential equation becomes

$$u(t) = 0 = RC \frac{dv_h(t)}{dt} + v_h(t) \qquad t < 0$$

You've already found the solution before time $t = 0$, because $v_h(t)$ is the solution to the homogeneous equation:

$$v_h(t) = c_1 e^{-\frac{t}{RC}}$$

You determine the arbitrary constant c_1 after finding the particular solution and applying the initial condition V_0 of 0 volts.

Now find the particular solution $v_p(t)$ when $u(t) = 1$ after $t = 0$.

After time $t = 0$, a unit step input describes the transient voltage behavior across the capacitor. The capacitor voltage reacting to a step input is called the *step response*.

For a step input $v_T(t) = u(t)$, you have a first-order differential equation:

$$u(t) = RC \frac{dv(t)}{dt} + v(t)$$

You already know that the value of the step $u(t)$ is equal to 1 after $t = 0$. Substitute $u(t) = 1$ into the preceding equation:

$$1 = RC \frac{dv_p(t)}{dt} + v_p(t) \qquad t \geq 0$$

Solve for the capacitor voltage $v_p(t)$, which is the particular solution. The particular solution always depends on the actual input signal.

Because the input is a constant after $t = 0$, the particular solution $v_p(t)$ is assumed to be a constant V_A as well.

The derivative of a constant is 0, which implies the following:

$$\frac{d}{dt}(V_A) = 0$$

Now substitute $v_p(t) = V_A$ and its derivative into the first-order differential equation:

$$1 = RC \underbrace{\frac{dv_p(t)}{dt}}_{=0} + \underbrace{v_p(t)}_{=V_A} \qquad \rightarrow \qquad v_p(t) = 1 = V_A \qquad t \geq 0$$

After a relatively long period of time, the particular solution follows the unit step input with strength $V_A = 1$. In general, a step input with strength V_A or $V_A u(t)$ leads to a capacitor voltage of V_A.

After finding the homogeneous and particular solutions, you add up the two solutions to get the zero-state response $v_{ZS}(t)$. You find c_1 by applying the initial condition that's equal to 0.

Adding up the homogeneous solution and the particular solution, you have $v_{ZS}(t)$:

$$v_{ZS}(t) = v_h(t) + v_p(t)$$

Substituting in the homogeneous and particular solutions gives you

$$v_{ZS}(t) = c_1 e^{-\left(\frac{t}{RC}\right)} + V_A$$

At $t = 0$, the initial condition is $v_c(0) = 0$ for the zero-state response. You now calculate $v_{ZS}(0)$ as

$$v_{ZS}(0) = c_1 e^{-\left(\frac{0}{RC}\right)} + V_A$$
$$0 = c_1 + V_A$$

Next, solve for c_1:

$$c_1 = -V_A$$

Substitute c_1 into the zero-state equation to produce the complete solution of the zero-state response $v_{ZS}(t)$:

$$v_{ZS}(t) = V_A\left(1 - e^{-\frac{t}{RC}}\right)$$

Adding the zero-input and zero-state responses to find the total response

You finally add up the zero-input response $v_{ZI}(t)$ and the zero-state response $v_{ZS}(t)$ to get the total response $v(t)$:

$$v(t) = v_{ZI}(t) + v_{ZS}(t)$$
$$= V_0 e^{-\left(\frac{t}{RC}\right)} + V_A\left(1 - e^{-\left(\frac{t}{RC}\right)}\right)$$

Time to verify whether the solution is reasonable. When $t = 0$, the initial voltage across the capacitor is

$$v(0) = V_0 e^{-\left(\frac{0}{RC}\right)} + V_A\left(1 - e^{-\left(\frac{0}{RC}\right)}\right)$$
$$= V_0$$

You bet this is a true statement! But you can check out when the initial conditions die out after a long period of time if you feel unsure about your solution. The output should just be related to the input voltage or step voltage.

After a long period of time (or after 5 time constants), you get the following:

$$v(\infty) = V_0 e^{-\left(\frac{\infty}{RC}\right)} + V_A\left(1 - e^{-\left(\frac{\infty}{RC}\right)}\right)$$
$$= V_A$$

Another true statement. The output voltage follows the step input with strength V_A after a long time. In other words, the capacitor voltage charges to a value equal to the strength V_A of the step input after the initial condition dies out in about 5 time constants.

Try this example with these values: V_0 = 5 volts, V_A = 10 volts, and RC = 1 second. You should get the capacitor voltage charging from an initial voltage of 5 volts and a final voltage of 10 volts after 5 seconds (5 time constants). Using the given values, you get the plot in Figure 13-4. The plot starts at 5 volts, and you end up at 10 volts after 5 time constants (5 seconds = $5RC$). So this example shows how changing voltage states takes time. Circuits with capacitors don't change voltages instantaneously. A large resistor also slows things down. That's why the time constant RC takes into account how the capacitor voltage will change from one voltage state to another.

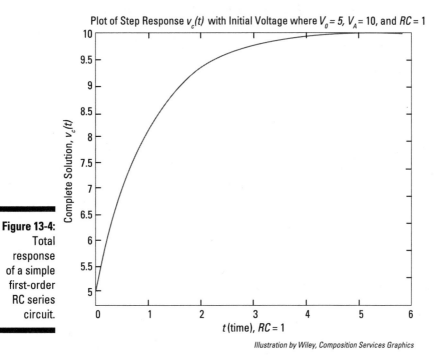

Plot of Step Response $v_c(t)$ with Initial Voltage where V_0 = 5, V_A = 10, and RC = 1

Figure 13-4:
Total response of a simple first-order RC series circuit.

The total capacitor voltage consists of the zero-input response and a zero-state response:

$$v(t) = \underbrace{5e^{-\frac{t}{RC}}}_{\substack{\text{zero-input} \\ \text{response}}} + \underbrace{10\left(1 - e^{-\left(\frac{t}{RC}\right)}\right)}_{\substack{\text{zero-state} \\ \text{response}}}$$

Illustration by Wiley, Composition Services Graphics

The RC time constant

The following table shows the various output values of the capacitor voltage of a homogeneous RC series circuit given by multiple time constants. After 5 time constants, the output voltage decays to less than 1 percent of the initial voltage V_0.

Time t	Solution v(t) $$v(t) = V_0 e^{-\left(\frac{t}{RC}\right)}$$	v(t) with Evaluation of the Natural Exponential Function
$t = 0$	$v(0) = V_0 e^{-\left(\frac{0}{RC}\right)} = V_0$	$v(0) = V_0$
$t = RC$	$v(RC) = V_0 e^{-\left(\frac{RC}{RC}\right)} = V_0 e^{-1}$	$v(RC) = 0.3679 V_0$
$t = 2RC$	$v(2RC) = V_0 e^{-\left(\frac{2RC}{RC}\right)} = V_0 e^{-2}$	$v(2RC) = 0.1353 V_0$
$t = 3RC$	$v(3RC) = V_0 e^{-\left(\frac{3RC}{RC}\right)} = V_0 e^{-3}$	$v(3RC) = 0.0498 V_0$
$t = 4RC$	$v(4RC) = V_0 e^{-\left(\frac{4RC}{RC}\right)} = V_0 e^{-4}$	$v(4RC) = 0.0183 V_0$
$t = 5RC$	$v(5RC) = V_0 e^{-\left(\frac{5RC}{RC}\right)} = V_0 e^{-5}$	$v(5RC) = 0.0067 V_0$

This equation shows that the total response is a combination of two outputs added together: one output due only to the initial voltage V_0 = 5 volts (at time t = 0) and the other due only to the step input with strength V_A = 10 volts (after time t = 0).

Analyzing a Parallel Circuit with a Single Resistor and Inductor

One type of first-order circuit consists of a resistor (or a network of resistors) and a single inductor. Analyzing such a parallel RL circuit, like the one in Figure 13-5, follows the same process I describe for analyzing an RC series circuit. I walk you through each of the steps in the following sections.

If your RL parallel circuit has an inductor connected with a network of resistors rather than a single resistor, you can use the same approach to analyze the circuit. But you have to find the Norton equivalent first, reducing the resistor network to a single resistor in parallel with a single current source. I cover the Norton approach in Chapter 8.

Figure 13-5:
A first-order
RL parallel
circuit.

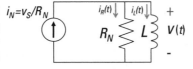

RL-parallel circuit

Starting with the simple RL parallel circuit

Because the resistor and inductor are connected in parallel in Figure 13-5, they must have the same voltage $v(t)$. The resistor current $i_R(t)$ is based on Ohm's law:

$$i_R(t) = \frac{v(t)}{R}$$

The element constraint for an inductor (see Chapter 12) is given as

$$v(t) = L\frac{di(t)}{dt}$$

where $i(t)$ is the inductor current and L is the inductance.

You need a changing current to generate voltage across an inductor. If the inductor current doesn't change, there's no inductor voltage, which implies a short circuit.

Now substitute $v(t) = Ldi(t)/dt$ into Ohm's law because you have the same voltage across the resistor and inductor:

$$i_R(t) = \left(\frac{L}{R}\right)\frac{di(t)}{dt}$$

Kirchhoff's current law (KCL) says the incoming currents are equal to the outgoing currents at a node. Use KCL at Node A of Figure 13-5 to get
$$i_N(t) = i_R(t) + i(t).$$

Substitute $i_R(t)$ into the KCL equation to give you

$$i_N(t) = \left(\frac{L}{R}\right)\frac{di(t)}{dt} + i(t)$$

The RL parallel circuit is a first-order circuit because it's described by a first-order differential equation, where the unknown variable is the inductor current $i(t)$. A circuit containing a single equivalent inductor and an equivalent resistor is a first-order circuit.

Knowing the inductor current gives you the magnetic energy stored in an inductor.

In general, the inductor current is referred to as a *state variable* because the inductor current describes the behavior of the circuit.

Calculating the zero-input response for an RL parallel circuit

Figure 13-6 shows how the RL parallel circuit is split up into two problems: the zero-input response and the zero-state response. This section starts off with the zero-input response, and the next section analyzes the zero-state response.

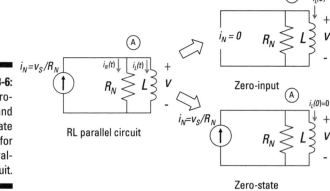

Figure 13-6: Zero-input and zero-state response for an RL parallel circuit.

To simplify matters, you set the input source (or forcing function) equal to 0: $i_N(t) = 0$ amps. This means no input current for all time — a big, fat zero. The first-order differential equation reduces to

$$i_N(t) = 0 = \left(\frac{L}{R}\right)\frac{di_{ZI}(t)}{dt} + i_{ZI}(t) \quad \text{or} \quad i(t) = -\left(\frac{L}{R}\right)\frac{di_{ZI}(t)}{dt}$$

For an input source of no current, the inductor current i_{ZI} is called a *zero-input response*. No external forces are acting on the circuit except for its initial state (or inductor current, in this case). The output is due to some initial inductor current I_0 at time $t = 0$.

You make a reasonable guess at the solution (the natural exponential function!) and substitute your guess into the RL first-order differential equation. Assume the inductor current and solution to be

$$i_{ZI}(t) = Be^{kt}$$

This is a reasonable guess because the time derivative of an exponential is also an exponential. Like a good friend, the exponential function won't let you down when solving these differential equations.

You determine the constants B and k next. Substitute your guess $i_{ZI}(t) = Be^{kt}$ into the differential equation:

$$i_{ZI}(t) = -\left(\frac{L}{R}\right)\frac{di_{ZI}(t)}{dt}$$

Replacing $i_{ZI}(t)$ with Be^{kt} and doing some math gives you the following:

$$Be^{kt} = -\left(\frac{L}{R}\right)\frac{d}{dt}\left(Be^{kt}\right)$$
$$= -\left(\frac{L}{R}\right)\left(kBe^{kt}\right)$$

You have the characteristic equation after factoring out Be^{kt}:

$$Be^{kt}\left[1 + \left(\frac{L}{R}\right)k\right] = 0$$

The characteristic equation gives you an algebraic problem to solve for the constant k:

$$1+\left(\frac{L}{R}\right)k = 0$$

$$k = -\frac{R}{L}$$

Use $k = -R/L$ and the initial inductor current I_0 at $t = 0$. This implies that $B = I_0$, so the zero-input response $i_{ZI}(t)$ gives you the following:

$$i_{ZI}(t) = I_0 e^{-\left(\frac{R}{L}\right)t}$$

The constant L/R is called the *time constant*. The time constant provides a measure of how long an inductor current takes to go to 0 or change from one state to another.

Calculating the zero-state response for an RL parallel circuit

Zero-state response means zero initial conditions. For the zero-state circuit in Figure 13-6, zero initial conditions means looking at the circuit with zero inductor current at $t < 0$. You need to find the homogeneous and particular solutions to get the zero-state response.

Next, you have zero initial conditions and an input current of $i_N(t) = u(t)$, where $u(t)$ is a unit step input.

When the step input $u(t) = 0$, the solution to the differential equation is the solution $i_h(t)$:

$$i_h(t) = c_1 e^{-\left(\frac{R}{L}\right)t}$$

The inductor current $i_h(t)$ is the solution to the homogeneous first-order differential equation:

$$u(t) = 0 = \left(\frac{L}{R}\right)\frac{di_h(t)}{dt} + i_h(t) \qquad t < 0$$

This solution is the general solution for the zero input. You find the constant c_1 after finding the particular solution and applying the initial condition of no inductor current.

After time $t = 0$, a unit step input describes the transient inductor current. The inductor current for this step input is called the *step response*. Not very creative, I know, but it does remind you of the step input.

You find the particular solution $i_p(t)$ by setting the step input $u(t)$ equal to 1. For a unit step input $i_N(t) = u(t)$, substitute $u(t) = 1$ into the differential equation:

$$u(t) = 1 = \left(\frac{L}{R}\right)\frac{di_p(t)}{dt} + i_p(t) \qquad t \geq 0$$

The particular solution $i_p(t)$ is the solution for the differential equation when the input is a unit step $u(t) = 1$ after $t = 0$.

Because $u(t) = 1$ (a constant) after time $t = 0$, assume a particular solution $i_p(t)$ is a constant I_A.

Because the derivative of a constant is 0, the following is true:

$$\frac{d}{dt}(I_A) = 0$$

Substitute $i_p(t) = I_A$ into the first-order differential equation:

$$1 = \left(\frac{R}{L}\right)\underbrace{\frac{di_p(t)}{dt}}_{=0} + \underbrace{i_p(t)}_{=I_A} \qquad \rightarrow \qquad i(t) = 1 = I_A \qquad t \geq 0$$

The particular solution eventually follows the form of the input because the zero-input (or free response) diminishes to 0 over time. You can generalize the result when the input step has strength I_A or $I_A u(t)$.

You need to add the homogeneous solution $i_h(t)$ and the particular solution $i_p(t)$ to get the zero-state response:

$$i_{ZS}(t) = i_h(t) + i_p(t)$$
$$i_{ZS}(t) = c_1 e^{-\left(\frac{R}{L}\right)t} + I_A$$

At $t = 0$, the initial condition is 0 because this is a zero-state calculation. To find c_1, apply $i_{ZS}(0) = 0$:

$$i_{ZS}(0) = c_1 e^{-\left(\frac{R}{L}\right)0} + I_A$$
$$0 = c_1 + I_A$$

Solving for c_1 gives you

$$c_1 = -I_A$$

Substituting c_1 into the zero-state response $i_{ZS}(t)$, you wind up with

$$i_{ZS}(t) = -I_A e^{-\left(\frac{R}{L}\right)t} + I_A$$
$$i_{ZS}(t) = I_A\left(1 - e^{-\left(\frac{R}{L}\right)t}\right)$$

Adding the zero-input and zero-state responses to find the total response

To obtain the total response for the RL parallel circuit, you need to add up the two solutions, the zero-input and zero-state responses:

$$i(t) = i_{ZI}(t) + i_{ZS}(t)$$

Substitute the zero-input and zero-state responses from the preceding sections into this equation, which gives you

$$i(t) = I_0 e^{-\left(\frac{R}{L}\right)t} + I_A\left(1 - e^{-\left(\frac{R}{L}\right)t}\right)$$

Check out the total response to verify the solution $i(t)$. When $t = 0$, the initial inductor current is

$$i(0) = I_0 e^{-\left(\frac{R}{L}\right)0} + I_A\left(1 - e^{-\left(\frac{R}{L}\right)0}\right)$$
$$i(0) = I_0$$

This is a true statement — for sure, for sure. If you're still not convinced, figure out when the initial condition dies out. The output should just be related to the input current or step current for this example.

After a long period of time (5 time constants), you get the following:

$$i(\infty) = I_0 e^{-\left(\frac{R}{L}\right)\infty} + I_A \left(1 - e^{-\left(\frac{R}{L}\right)\infty}\right)$$

$$i(\infty) = 0 + I_A(1-0)$$

$$i(\infty) = I_A$$

The output inductor current is just the step input having a strength of I_A. In other words, the inductor current reaches a value equal to the step input's strength I_A after the initial condition dies out in about 5 time constants of L/R, or $5L/R$. You see inductor currents don't change instantaneously. With inductors, currents change gradually in going from one state to another. A parallel resistor slows things down. That's why the time constant L/R takes into account how fast inductor currents change from one state to another.

The complete response of the inductor current follows the same shape of the capacitor voltage in Figure 13-4. The shape starts at some initial current and goes to another current state after 5 time constants.

The *L/R* time constant

For zero-input and initial current I_0, the output inductor current for a parallel RL circuit is

$$i(t) = I_0 e^{-\left(\frac{R}{L}\right)t}$$

The time constant is $t = L/R$. After 5 time constants, the output inductor current decays to less than 1 percent of the initial current I_0. The inductor current follows the same shape as the capacitor voltage given in Figure 13-3.

Chapter 14

Analyzing Second-Order Circuits

. . . .

In This Chapter

▶ Focusing on second-order differential equations

▶ Analyzing an RLC series circuit

▶ Analyzing an RLC parallel circuit

. . . .

Second-order circuits consist of capacitors, inductors, and resistors. In math terms, circuits that have both an inductor and a capacitor are described by second-order differential equations — hence the name *second-order circuits*. This chapter clues you in to what's unique about analyzing second-order circuits and then walks you through the analysis of an RLC (resistor, inductor, capacitor) series circuit and an RLC parallel circuit.

For a refresher on second-order differential equations, refer to your textbook or *Differential Equations For Dummies* by Steven Holzner (Wiley).

Examining Second-Order Differential Equations with Constant Coefficients

If you can use a second-order differential equation to describe the circuit you're looking at, then you're dealing with a second-order circuit. Circuits that include an inductor, capacitor, and resistor connected in series or in parallel are second-order circuits. Figure 14-1 shows second-order circuits driven by an input source, or forcing function.

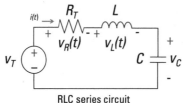
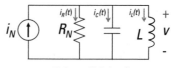

Figure 14-1:
Examples
of second-
order
circuits.

RLC series circuit   RLC parallel circuit

Illustration by Wiley, Composition Services Graphics

Getting a unique solution to a second-order differential equation requires
knowing the initial states of the circuit. For a second-order circuit, you need
to know the initial capacitor voltage and the initial inductor current. Knowing
these states at time $t = 0$ provides you with a unique solution for all time after
time $t = 0$.

REMEMBER

Use these steps when solving a second-order differential equation for a
second-order circuit:

1. **Find the zero-input response by setting the input source to 0, such that
   the output is due only to initial conditions.**

2. **Find the zero-state response by setting the initial conditions equal to
   0, such that the output is due only to the input signal.**

   Zero initial conditions means you have 0 initial capacitor voltage and 0
   initial inductor current.

   The zero-state response requires you to find the homogeneous and particular solutions:

   • **Homogeneous solution:** When there's no input signal or forcing
     function — that is, when $v_T(t) = 0$ or $i_N(t) = 0$ — you have the
     homogeneous solution.

   • **Particular solution:** When you have a nonzero input, the solution
     follows the form of the input signal, giving you the *particular solu-
     tion.* For example, if your input is a constant, then your particular
     solution is also a constant. Likewise, if you have a sine or cosine
     function as an input, then the output is a combination of sine and
     cosine functions.

3. **Add up the zero-input and zero-state responses to get the total
   response.**

   Because you're dealing with linear circuits, you want to use superposition to find the total response. I show you the superposition technique
   in Chapter 7.

In the following sections, I show you how to find the total response for a second-order differential equation with constant coefficients. I first find the homogeneous solution by using an algebraic characteristic equation and assuming the solutions are exponential functions. The roots to the characteristic equation give you the constants found in the exponent of the exponential function.

Later in this chapter, I analyze an RLC series circuit by applying the preceding steps to get the total response. I set up the appropriate equations using Kirchhoff's voltage law (KVL) and device equations for a capacitor and inductor. Then I determine the zero-input response followed by the calculation of the zero-state response. Finally, I analyze an RLC parallel circuit using the concept of duality, which replaces quantities with their dual quantities. The resulting equations for an RLC parallel circuit are similar to the equations for an RLC series circuit.

Guessing at the elementary solutions: The natural exponential function

I'm giving you just one approach to solving second-order circuits. The good news is that it converts a problem involving a differential equation to one that uses only algebra.

Consider the following differential equation as a numerical example with zero forcing function $v_T(t) = 0$:

$$\frac{d^2v}{dt^2} + 5\frac{dv}{dt} + 6v = 0$$

The solution to this differential equation is called the *homogeneous solution* $v(t)$. One classic approach entails giving your best shot at guessing the solution. Try $v(t) = e^{kt}$. The exponential function works for a first-order equation, so it should work for a second-order equation, too. When you take the derivative of the natural exponential e^{kt}, you get the same thing multiplied by some constant k. You see how the exponential function is your true amigo in solving differential equations like this.

From calculus to algebra: Using the characteristic equation

To solve a homogeneous differential equation, you can convert the differential equation into a characteristic equation, which you solve using algebra. You do this by substituting your guess $v(t) = e^{kt}$ (from the preceding section) into the homogeneous differential equation:

$$\frac{d^2v}{dt^2} + 5\frac{dv}{dt} + 6v = 0$$

$$\frac{d^2}{dt^2}\left(e^{kt}\right) + 5\frac{d}{dt}\left(e^{kt}\right) + 6e^{kt} = 0$$

Factoring out e^{kt} leads you to a characteristic equation:

$$\left(k^2 + 5k + 6\right)e^{kt} = 0$$

The coefficient of e^{kt} must be 0, so you can solve for k as follows:

$$k^2 + 5k + 6 = 0$$
$$k = -2, -3$$

Setting the algebraic equation to 0 gives you a *characteristic equation*. The constant roots –2 and –3 determine the features of the solution $v(t)$.

From these roots, you get a homogeneous solution that's a combination of the solutions e^{-2t} and e^{-3t}:

$$v(t) = c_1 e^{-2t} + c_2 e^{-3t}$$

The constants c_1 and c_2 are determined by the initial conditions when $t = 0$.

Analyzing an RLC Series Circuit

One second-order circuit consists of a resistor (or network of resistors) hooked up in series with both a capacitor and an inductor. The left diagram in Figure 14-2 shows you what such an RLC series circuit looks like. The rest of Figure 14-2 shows you the RLC series circuit broken into two circuits: One deals with the initial condition, and the other deals with the input source. The top-right diagram shows the zero-input response, setting the input to 0, and the bottom-right diagram deals with the zero-state response, setting the initial conditions to 0.

The following sections walk you through the analysis process for an RLC series circuit.

Figure 14-2:
A second-order RLC series circuit broken into circuits to help you find the zero-input response and zero-state response.

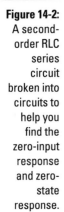 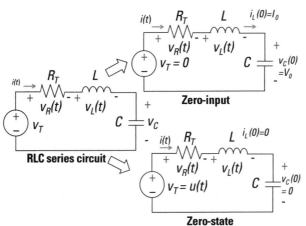

Zero-input

RLC series circuit

Zero-state

Illustration by Wiley, Composition Services Graphics

Setting up a typical RLC series circuit

The simple RLC series circuit in Figure 14-2 is driven by a voltage source. Kirchhoff's voltage law (KVL) says the sum of the voltage drops and rises around a loop of a circuit is equal to 0. Using KVL for this circuit gives you the following:

$$v_T(t) = v_R(t) + v_L(t) + v_C(t)$$

The subscript letter R is for the resistor, L is for the inductor, and C is for the capacitor — pretty straightforward, huh? Because these devices are connected in series, they have the same current $i(t)$.

Next, you want to put the resistor voltage and the inductor voltage in terms of the capacitor voltage or its derivative. The voltage across the resistor $v_R(t)$ uses Ohm's law:

$$v_R(t) = Ri(t)$$

The element constraint for an inductor voltage $v_L(t)$ is

$$v_L(t) = L\frac{di(t)}{dt}$$

And for a capacitor current $i(t)$, the device constraint is

$$i(t) = C\frac{dv_c(t)}{dt}$$

Because the series current $i(t)$ flows through each device, you can substitute the capacitor current into the equations for the resistor voltage and inductor voltage. First substitute the capacitor current into the inductor voltage equation:

$$v_L(t) = L\frac{di(t)}{dt}$$

$$v_L(t) = LC\frac{d^2v_C(t)}{dt^2}$$

Next, plug the capacitor current $i(t) = Cdv_C(t)/dt$ into Ohm's law to get the resistor voltage $v_R(t)$:

$$v_R(t) = RC\frac{dv_C(t)}{dt}$$

Now you can plug $v_R(t)$ and $v_L(t)$ into the KVL equation, giving you all the device voltages in terms of the capacitor voltage (or its derivatives):

$$v_T(t) = LC\frac{d^2v_C}{dt^2} + RC\frac{dv_C(t)}{dt} + v_C(t)$$

You now have a second-order differential equation where the unknown function is the capacitor voltage $v_C(t)$. Knowing $v_C(t)$ gives you the electrical energy stored in the capacitor or the capacitor's state of charge.

In general, the capacitor voltage and inductor current are referred to as *state variables* because these quantities describe the behavior of the circuit at any time.

The RLC series circuit is a second-order circuit because it has two energy-storage devices. It can be described by a second-order differential equation having an applied input voltage $v_T(t)$.

Determining the zero-input response

The top-right diagram of Figure 14-2 shows the input signal $v_T(t) = 0$, which gives you the zero-input response. With the zero-input response, you have no input voltage for all time. This response comes from initial capacitor voltage V_0 and initial inductor current I_0 at time $t = 0$. With zero input, the second-order differential equation reduces to

$$v_T(t) = 0 = LC \frac{d^2 v_{ZI}(t)}{dt^2} + RC \frac{dv_{ZI}(t)}{dt} + v_{ZI}(t)$$

The capacitor voltage is called a *zero-input response*, $v_{ZI}(t)$. The top-right diagram of Figure 14-2 shows the input source set to 0 volts. No external forces (a battery, for example) are acting on the circuit except for its initial states, expressed by the capacitor voltage and inductor current.

You make a reasonable guess at the solution to $v_{ZI}(t)$: the exponential function (see the earlier section "Guessing at the elementary solutions: The natural exponential function"). Substitute the guess into the RLC second-order circuit equation. You can check and verify the solution afterward.

Assume the capacitor voltage and solution to be

$$v_{ZI}(t) = Ae^{kt}$$

The A and k are arbitrary constants of the zero-input response. You try an exponential function because the time derivative of an exponential is also an exponential.

Substitute the solution $v_{ZI}(t) = Ae^{kt}$ into the differential equation and simplify:

$$LC \frac{d^2 v_C(t)}{dt^2} + RC \frac{dv_C(t)}{dt} + v_C(t) = 0$$

$$LC \frac{d^2}{dt^2}(Ae^{kt}) + RC \frac{d}{dt}(Ae^{kt}) + Ae^{kt} = 0$$

$$LCk^2 Ae^{kt} + RCkAe^{kt} + Ae^{kt} = 0$$

Factoring out Ae^{kt} gives you the algebraic characteristic equation (I also factor out LC so that the leading coefficient on k^2 is 1):

$$Ae^{kt}\left(LCk^2 + RCk + 1\right) = 0$$

$$LC\left(Ae^{kt}\right)\left(k^2 + \frac{R}{L}k + \frac{1}{LC}\right) = 0$$

You've transformed the differential equation into an algebraic one. The coefficient of e^{kt} has to equal 0, so use that info to solve for the constant k:

$$k^2 + \frac{R}{L}k + \frac{1}{LC} = 0$$

$$k_1, k_2 = -\frac{R}{2L} \pm \frac{1}{2}\sqrt{\left(\frac{R}{L}\right)^2 - \frac{4}{LC}}$$

You now have three possible cases and roots under the radical:

Case 1: $\left[\left(\frac{R}{L}\right)^2 - \frac{4}{LC}\right] > 0 \;\rightarrow\;$ two different real roots $(\alpha_1 \text{ and } \alpha_2)$

Case 2: $\left[\left(\frac{R}{L}\right)^2 - \frac{4}{LC}\right] = 0 \;\rightarrow\;$ two equal real roots $(\alpha_1 = \alpha_2)$

Case 3: $\left[\left(\frac{R}{L}\right)^2 - \frac{4}{LC}\right] < 0 \;\rightarrow\;$ two complex conjugate roots $(\alpha \pm j\beta)$

Loosely speaking, Case 1 implies a large resistor R losing lots of energy as heat, with the initial states eventually dying out. Case 3 implies that the resistor is small, where stored energy is being exchanged between the capacitor and inductor. With a sinusoidal input, the stored energy switches between the electrical energy in the capacitor and the magnetic energy in the inductor. This back-and-forth sloshing of energy causes oscillations at the output. Case 2, with two equal and real roots, falls between these two behaviors. Case 2 achieves a faster response than Case 1 but doesn't suffer from the oscillations found in Case 3.

Later in this chapter, Figure 14-3 illustrates the effects of decreasing resistance when you have zero-input response, and Figure 14-4 illustrates the effects of decreasing resistance when you have a step response. Without getting into the analytical detail of how you arrive at these curves, you should observe that for decreasing resistance, you have increasing amplitude of oscillations.

When you know k_1 and k_2, you have the zero-input response $v_{ZI}(t)$. The response $v_{ZI}(t)$ comes from a combination of the two solutions:

$$v_{ZI}(t) = c_1 e^{k_1 t} + c_2 e^{k_2 t}$$

You find the constants c_1 and c_2 by applying two initial conditions: inductor current $i_L(0) = I_0$ and capacitor voltage $v_C(0) = V_0$. At time $t = 0$, the initial capacitor voltage is V_0, and you have

$$v_{ZI}(t) = v_C(t) = c_1 + c_2 = V_0$$

The same current flows through the inductor and the capacitor. You find the inductor's initial current based on the initial condition of the first derivative of capacitor voltage:

$$\frac{dv_C(t)}{dt}\bigg|_{t=0} = \frac{dv_C(0)}{dt} = \frac{i(0)}{C} = \frac{I_0}{C}$$

Taking the derivative of $v_C(t) = v_{ZI}(t)$ gives you the following:

$$\frac{dv_{ZI}(0)}{dt}\bigg|_{t=0} = k_1 c_1 + k_2 c_2 = \frac{I_0}{C}$$

Apply the two initial conditions to give you two equations having two unknowns, c_1 and c_2:

$$v_C(0) = V_0 \quad \rightarrow \quad c_1 + c_2 = V_0$$

$$\frac{dv_C(0)}{dt} = \frac{I_0}{C} \quad \rightarrow \quad k_1 c_1 + k_2 c_2 = \frac{I_0}{C}$$

Then solve for c_1 and c_2:

$$c_1 = \frac{k_2 V_0 - I_0 / C}{k_2 - k_1}$$

$$c_2 = \frac{-k_1 V_0 + I_0 / C}{k_2 - k_1}$$

The roots k_1 and k_2 of the characteristic equation reveal the form of the zero-input response $v_C(t) = v_{ZI}(t)$. Based on these roots, you have different solutions for the capacitor voltage $v_C(t)$, which is the *zero-input response* $v_{ZI}(t)$:

Case 1: $v_C(t) = c_1 e^{\alpha_1 t} + c_2 e^{\alpha_2 t}$ → two different real roots $(\alpha_1$ and $\alpha_2)$

Case 2: $v_C(t) = c_1 e^{\alpha t} + c_2 t e^{\alpha t}$ → two equal real roots $(\alpha$ and $\alpha)$

Case 3: $v_C(t) = c_1 e^{\alpha t} \cos \beta t + c_2 e^{\alpha t} \sin \beta t$ → complex roots $(\alpha \pm j\beta)$

Figure 14-3 shows three zero-input responses for various values of resistors. Note that real roots mean damping exponentials, and complex roots indicate oscillations. Case 1 (overdamped) doesn't give you oscillations, but the initial conditions die out the most slowly. For Case 3 (underdamped), you get to the desired state faster, but you have oscillations. Case 2 (critically damped) falls between Cases 1 and 3, with faster response than Case 1 and little or none of the oscillations found in Case 3.

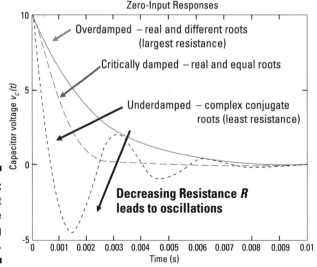

Figure 14-3:
Zero-input
response
for varying
resistance.

Illustration by Wiley, Composition Services Graphics

Calculating the zero-state response

Zero-state response means the response of a system under zero initial conditions, implying 0 capacitor voltage and 0 inductor current. When there's an input source $v_T(t)$, you need to find the solution to the homogeneous differential equation and the solution to the differential equation for a particular input.

The circuit in the bottom-right diagram of Figure 14-2 has zero initial conditions and an input voltage of $v_T(t) = u(t)$, where $u(t)$ is a unit step input (I introduce unit step functions in Chapter 11). Mathematically, you can describe a step function $u(t)$ as

$$u(t) = \begin{cases} 0 & t < 0 \\ 1 & t \geq 0 \end{cases}$$

The input signal is divided into two time intervals. When $t < 0$, $u(t) = 0$. In terms of the capacitor voltage $v(t)$, the second-order differential equation becomes

$$LC\frac{d^2v(t)}{dt^2} + RC\frac{dv(t)}{dt} + v(t) = 0 \qquad t < 0$$

Where $u(t) = 0$ for before time $t = 0$, you have the homogeneous solution $v_h(t)$ when the input is 0:

$$v_h(t) = C_1 e^{k_1 t} + C_2 e^{k_2 t}$$

You determine the arbitrary constants C_1 and C_2 after finding the particular solution and applying the initial condition V_0 of 0 volts. You find the particular solution $v_p(t)$ when $u(t) = 1$ after $t = 0$.

For a step input $v_T(t) = u(t)$, you have the following second-order differential equation:

$$u(t) = LC\frac{d^2v(t)}{dt^2} + RC\frac{dv(t)}{dt} + v(t) \qquad t \geq 0$$

After $t = 0$, the value of the step input $u(t)$ is equal to 1. Substitute $u(t) = 1$ into the preceding equation:

$$1 = LC\frac{d^2v_p(t)}{dt^2} + RC\frac{dv_p(t)}{dt} + v_p(t) \qquad t \geq 0$$

Solve for the capacitor voltage $v_p(t)$ to get the *particular solution,* or *forced response.* The particular solution always depends on the actual input signal.

Because the input is constant after $t = 0$, the particular solution $v_p(t)$ is assumed to be a constant, V_A, as well.

The derivative of a constant is 0:

$$\frac{d}{dt}(V_A) = 0$$

Substitute $v_p(t) = V_A$ and its derivative into the second-order differential equation:

$$1 = LC\underbrace{\frac{d^2v_p(t)}{dt^2}}_{=0} + RC\underbrace{\frac{dv_p(t)}{dt}}_{0} + v_p(t) \qquad \rightarrow \qquad v_p(t) = 1 = V_A \qquad t \geq 0$$

The particular solution eventually follows the step input after a relatively long period of time. In general, a step input with strength V_A or $V_A u(t)$ leads to a capacitor voltage of V_A.

After finding the solution due to the homogeneous differential equation and the solution for a particular input, you add up the two solutions to get the zero-state response $v_{ZS}(t)$. You find C_1 by applying the zero initial condition.

Adding up the two solutions gives you the zero-state response $v_{ZS}(t)$:

$$v_{ZS}(t) = v_h(t) + v_p(t)$$

Substituting the two solutions into this equation gives you the following:

$$v_{ZS}(t) = C_1 e^{k_1 t} + C_2 e^{k_2 t} + V_A$$
$$\frac{dv_{ZS}(t)}{dt} = C_1 k_1 e^{k_1 t} + C_2 k_2 e^{k_2 t}$$

By definition, at $t = 0$, the initial conditions for a circuit in a zero state is $v_C(0) = i_L(0) = 0$. The zero-state response $v_{ZS}(0)$ is

$$v_{ZS}(0) = 0 = C_1 + C_2 + V_A \qquad \rightarrow \qquad -V_A = C_1 + C_2$$
$$\frac{dv_{ZS}(0)}{dt} = 0 = C_1 k_1 + C_2 k_2$$

Solve for C_1 and C_2:

$$C_1 = \frac{k_2 V_A}{k_1 - k_2} \quad \text{and} \quad C_2 = \frac{k_1 V_A}{k_2 - k_1}$$

Finishing up with the total response

Add up the zero-input response $v_{ZI}(t)$ and the zero-state response $v_{ZS}(t)$ to get the total response $v(t)$:

$$v(t) = v_{ZI}(t) + v_{ZS}(t)$$
$$v(t) = c_1 e^{k_1 t} + c_2 e^{k_2 t} + C_1 e^{k_1 t} + C_2 e^{k_2 t} + V_A$$

Do you get good vibes from this solution? If not, you need to verify that the solution is reasonable. When $t = 0$, the initial voltage across the capacitor is

$$v(0) = c_1 e^{-k_1 \cdot 0} + c_2 e^{-k_2 \cdot 0} + C_1 e^{-k_1 \cdot 0} + C_2 e^{-k_2 \cdot 0} + V_A$$
$$= V_0$$

You can substitute c_1, c_2, C_1, and C_2 into this equation (based on the analysis in the previous sections) to confirm that this statement is true.

Next, check out the initial inductor current when you take the derivative of $v(t)$ and evaluate the derivative at $t = 0$:

$$\left. \frac{dv(0)}{dt} \right|_{t=0} = \frac{dv(0)}{dt} = c_1 k_1 e^{-k_1 \cdot 0} + c_2 k_2 e^{-k_2 \cdot 0} + C_1 k_1 e^{-k_1 \cdot 0} + C_2 k_2 e^{-k_2 \cdot 0}$$
$$= \frac{I_0}{C}$$

That's another true statement. If you're still not feeling good about your solution, look at when the initial conditions die out after a long period of time. The output should just be the step voltage. After a long period of time (or after 5 time constants), you get the following:

$$v(\infty) = c_1 e^{-k_1 \cdot \infty} + c_2 e^{-k_2 \cdot \infty} + C_1 e^{-k_1 \cdot \infty} + C_2 e^{-k_2 \cdot \infty} + V_A$$
$$= V_A$$

Another true statement! The output voltage follows the step input with strength V_A after an extended time. In other words, the capacitor voltage is equal to the strength V_A of the step input after the initial conditions die out.

Figure 14-4 shows several step responses for zero initial conditions for various values of decreasing resistance R. In this example, the step input has a strength of 10 volts. See how all the step responses end up at 10 volts after the time-varying output dies out. Although you initially reach the final value faster with decreasing resistance, you may end up with undesirable wavy behavior.

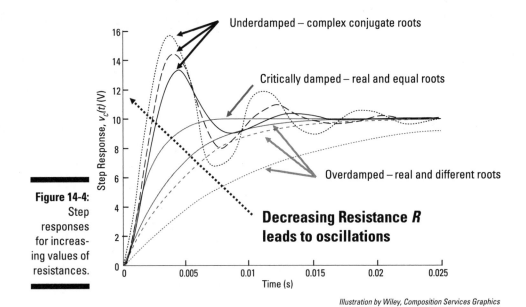

Figure 14-4:
Step
responses
for increas-
ing values of
resistances.

Analyzing an RLC Parallel Circuit Using Duality

One type of second-order circuit has a resistor, inductor, and capacitor connected in parallel. Check out the example RLC parallel circuit in Figure 14-5.
To analyze this second-order circuit, you use basically the same process as
for analyzing an RLC series circuit (see the preceding sections).

The left diagram of Figure 14-5 shows an input i_N with initial inductor current
I_0 and capacitor voltage V_0. The top-right diagram shows the input current
source i_N set equal to zero, which lets you solve for the zero-input response.
The bottom-right diagram shows the initial conditions (I_0 and V_0) set equal to
zero, which lets you obtain the zero-state response.

In the following sections, I show you how you can use the concept of duality
to obtain results similar to the ones you find in an RLC series circuit. With
duality, you substitute every electrical term in an equation with its dual, or
counterpart, and get another correct equation. For example, voltage and current are dual variables.

Figure 14-5:
A second-order RLC parallel circuit broken into circuits to help you find the zero-input response and zero-state response.

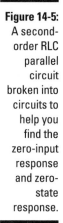

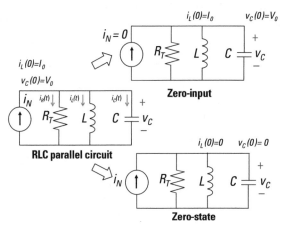

Illustration by Wiley, Composition Services Graphics

Setting up a typical RLC parallel circuit

Because the components of the circuit in Figure 14-5 are connected in parallel, you set up the second-order differential equation by using Kirchhoff's current law (KCL). KCL says the sum of the incoming currents equals the sum of the outgoing currents at a node. Using KCL at Node A of Figure 14-5 gives you

$$i_N(t) = i_R(t) + i_C(t) + i_L(t)$$

Next, put the resistor current and capacitor current in terms of the inductor current. The resistor current $i_R(t)$ is based on the old, reliable Ohm's law:

$$i_R(t) = \frac{v(t)}{R}$$

The element constraint for an inductor is given as

$$v(t) = L\frac{di(t)}{dt}$$

The current $i_L(t)$ is the inductor current, and L is the inductance. This constraint means a changing current generates an inductor voltage. If the inductor current doesn't change, there's no inductor voltage, implying a short circuit.

Parallel devices have the same voltage $v(t)$. You use the inductor voltage $v(t)$ that's equal to the capacitor voltage to get the capacitor current $i_C(t)$:

$$i_C(t) = C \frac{dv(t)}{dt} = LC \frac{d^2 i_L(t)}{dt^2}$$

Now substitute $v(t) = L di_L(t)/dt$ into Ohm's law, because you also have the same voltage across the resistor and inductor:

$$i_R(t) = \left(\frac{L}{R}\right) \frac{di_L(t)}{dt}$$

Substitute the values of $i_R(t)$ and $i_C(t)$ into the KCL equation to give you the device currents in terms of the inductor current:

$$i_N(t) = LC \frac{d^2 i_L(t)}{dt^2} + \left(\frac{L}{R}\right) \frac{di_L(t)}{dt} + i_L(t)$$

The RLC parallel circuit is described by a second-order differential equation, so the circuit is a second-order circuit. The unknown is the inductor current $i_L(t)$.

The analysis of the RLC parallel circuit follows along the same lines as the RLC series circuit. Compare the preceding equation with the second-order equation derived from the RLC series circuit (see the earlier section "Calculating the zero-state response" for details):

$$v_T(t) = LC \frac{d^2 v_C(t)}{dt^2} + RC \frac{dv_C(t)}{dt} + v_C(t)$$

The two differential equations have the same form. The unknown solution for the parallel RLC circuit is the inductor current, and the unknown for the series RLC circuit is the capacitor voltage. These unknowns are dual variables.

With *duality*, you can replace every electrical term in an equation with its dual and get another correct equation. If you use the following substitution of variables in the differential equation for the RLC series circuit, you get the differential equation for the RLC parallel circuit.

$$v_C(t) \leftrightarrow i_L(t)$$
$$v_T(t) \leftrightarrow i_N(t)$$
$$L \leftrightarrow C$$
$$C \leftrightarrow L$$
$$R \leftrightarrow G = 1/R$$
$$\text{series} \leftrightarrow \text{parallel}$$
$$\text{KVL} \leftrightarrow \text{KCL}$$

Duality allows you to simplify your analysis when you know prior results. Yippee!

Finding the zero-input response

The results you obtain for an RLC parallel circuit are similar to the ones you get for the RLC series circuit (I cover that series circuit earlier in "Analyzing an RLC Series Circuit").

As shown in the earlier section "Guessing at the elementary solutions: The natural exponential function," you have a characteristic equation to the homogeneous equation. For a parallel circuit, you have a second-order and homogeneous differential equation given in terms of the inductor current:

$$0 = \frac{d^2 i_L}{dt^2} + \left(\frac{1}{RC}\right)\frac{di_L(t)}{dt} + \left(\frac{1}{LC}\right)i_L(t)$$

$$k^2 + \frac{1}{RC}k + \frac{1}{LC} = 0 \qquad \rightarrow \qquad k_1, k_2 = -\frac{1}{2RC} \pm \frac{1}{2}\sqrt{\left(\frac{1}{RC}\right)^2 - \frac{4}{LC}}$$

The preceding equation gives you three possible cases under the radical:

Case 1: $\left[\left(\frac{1}{RC}\right)^2 - \frac{4}{LC}\right] > 0 \rightarrow$ two different real roots (α_1 and α_2)

Case 2: $\left[\left(\frac{1}{RC}\right)^2 - \frac{4}{LC}\right] = 0 \rightarrow$ two equal real roots ($\alpha_1 = \alpha_2$)

Case 3: $\left[\left(\frac{1}{RC}\right)^2 - \frac{4}{LC}\right] < 0 \rightarrow$ two complex conjugate roots ($\alpha \pm j\beta$)

The zero-input responses of the inductor responses resemble the form in Figure 14-3, which describes the capacitor voltage.

When you have k_1 and k_2, you have the zero-input response $i_{ZI}(t)$. The solution gives you

$$i_{ZI}(t) = c_1 e^{k_1 t} + c_2 e^{k_2 t}$$

You can find the constants c_1 and c_2 by using the results found in the RLC series circuit, which are given as

$$c_1 = \frac{k_2 V_0 - I_0 / C}{k_2 - k_1}$$

$$c_2 = \frac{-k_1 V_0 + I_0 / C}{k_2 - k_1}$$

Apply duality to the preceding equation by replacing the voltage, current, and inductance with their duals (current, voltage, and capacitance) to get c_1 and c_2 for the RLC parallel circuit:

$$c_1 = \frac{k_2 I_0 - V_0 / L}{k_2 - k_1}$$

$$c_2 = \frac{-k_1 I_0 + V_0 / L}{k_2 - k_1}$$

After you plug in the dual variables, finding the constants c_1 and c_2 is easy.

Arriving at the zero-state response

Zero-state response means zero initial conditions. You need to find the homogeneous and particular solutions of the inductor current when there's an input source $i_N(t)$. Zero initial conditions means looking at the circuit when there's 0 inductor current and 0 capacitor voltage.

When $t < 0$, $u(t) = 0$. The second-order differential equation becomes the following, where $i_L(t)$ is the inductor current:

$$i_N(t) = LC \frac{d^2 i_L(t)}{dt^2} + \left(\frac{L}{R}\right) \frac{di_L(t)}{dt} + i_L(t) \qquad t < 0$$

For a step input where $u(t) = 0$ before time $t = 0$, the homogeneous solution $i_h(t)$ is

$$i_h(t) = C_1 e^{k_1 t} + C_2 e^{k_2 t}$$

Adding the homogeneous solution to the particular solution for a step input $I_A u(t)$ gives you the zero-state response $i_{ZS}(t)$:

$$i_{ZS}(t) = i_h(t) + \underbrace{i_p(t)}_{I_A}$$

Now plug in the values of $i_h(t)$ and $i_p(t)$:

$$i_{ZS}(t) = C_1 e^{k_1 t} + C_2 e^{k_2 t} + I_A$$

Here are the results of C_1 and C_2 for the RLC series circuit:

$$C_1 = \frac{k_2 V_A}{k_1 - k_2} \quad \text{and} \quad C_2 = \frac{k_1 V_A}{k_2 - k_1}$$

You now apply duality through a simple substitution of terms in order to get C_1 and C_2 for the RLC parallel circuit:

$$C_1 = \frac{k_2 I_A}{k_1 - k_2} \quad \text{and} \quad C_2 = \frac{k_1 I_A}{k_2 - k_1}$$

Getting the total response

You finally add up the zero-input response $i_{ZI}(t)$ and the zero-state response $i_{ZS}(t)$ to get the total response $i_L(t)$:

$$i_L(t) = i_{ZI}(t) + i_{ZS}(t)$$
$$i_L(t) = c_1 e^{k_1 t} + c_2 e^{k_2 t} + C_1 e^{k_1 t} + C_2 e^{k_2 t} + I_A$$

The solution resembles the results for the RLC series circuit. Also, the step responses of the inductor current follow the same form as the ones shown in the step responses found in Figure 14-4 for the capacitor voltage.

Part V
Advanced Techniques and Applications in Circuit Analysis

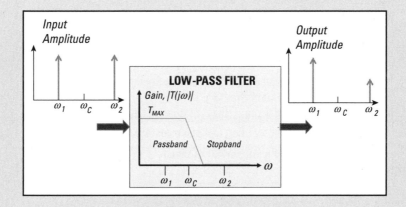

Design a filter to improve the sound quality of a speaker system at www.dummies.com/extras/circuitanalysis.

In this part . . .

✔ Use phasors to describe a sinusoidal signal.

✔ Transform functions using the Laplace technique so you can solve problems algebraically.

✔ Analyze circuits that have voltage and current signals that change with time by using Laplace transforms.

✔ Explore filters and frequency response.

Chapter 15

Phasing in Phasors for Wave Functions

. . . .

In This Chapter

▶ Describing circuit behavior with phasors

▶ Mixing phasors with impedance and Ohm's law

▶ Applying phasor techniques to circuits

. . . .

*P*hasors — not to be confused with the phasers from *Star Trek* — are rotating vectors you can use to describe the behavior of circuits that include capacitors and inductors. Phasors make the analysis of such circuits easier because instead of dealing with differential equations, you just have to work with complex numbers. I don't know about you, but I'd take working with complex numbers to solve circuits any day over using differential equations.

Phasor analysis applies when your input is a sine wave (or sinusoidal signal). A phasor contains information about the amplitude and phase of the sinusoidal signal. Frequency isn't part of phasor form because the frequency doesn't change in a linear circuit.

This chapter introduces phasors and explains how they represent a circuit's *i-v* characteristics. I then show you how phasors let you summarize the complex interactions among resistors, capacitors, and inductors as a tidy value called *impedance*. Finally, you see how phasors let you analyze circuits with storage devices algebraically, in the same way you analyze circuits with only resistors.

Taking a More Imaginative Turn with Phasors

A phasor is a complex number in polar form. When you plot the amplitude and phase shift of a sinusoid in a complex plane, you form a phase vector, or phasor.

As I'm sure you're well aware from algebra class, a complex number consists of a real part and an imaginary part. For circuit analysis, think of the real part as tying in with resistors that get rid of energy as heat and the imaginary part as relating to stored energy, like the kind found in inductors and capacitors.

You can also think of a phasor as a rotating vector. Unlike a vector having magnitude and direction, a phasor has *magnitude* V_A and *angular displacement* ϕ. You measure angular displacement in the counterclockwise direction from the positive *x*-axis.

Figure 15-1 shows a diagram of a voltage phasor as a rotating vector at some frequency, with its tail at the origin. If you need to add or subtract phasors, you can convert the vector into its *x*-component ($V_A \cos \phi$) and its *y*-component ($V_A \sin \phi$) with some trigonometry.

Figure 15-1:
A phasor is
a rotating
vector in
the complex
plane.

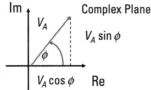

Illustration by Wiley, Composition Services Graphics

The following sections explain how to find the different forms of phasors and introduce you to the properties of phasors.

Finding phasor forms

Phasors, which you describe with complex numbers, embody the amplitude and phase of a sinusoidal voltage or current. The phase is the angular shift of the sinusoid, which corresponds to a time shift t_0. So if you have $\cos[\omega(t - t_0)]$, then $\omega t_0 = \phi_0$, where ϕ_0 is the angular phase shift.

To establish a connection between complex numbers and sine and cosine waves, you need the complex exponential $e^{j\theta}$ and Euler's formula:

$$e^{j\theta} = \cos\theta + j\sin\theta$$

where $j = \sqrt{-1}$.

The left side of Euler's formula is the *polar* phasor form, and the right side is the *rectangular* phasor form. You can write the cosine and sine as follows:

$$\cos\theta = \mathrm{Re}\left[\, e^{j\theta} \,\right]$$
$$\sin\theta = \mathrm{Im}\left[\, e^{j\theta} \,\right]$$

Re[ ] denotes the real part of a complex number, and Im[ ] denotes the imaginary part of a complex number.

Figure 15-2 shows a cosine function and a shifted cosine function with a phase shift of $\pi/2$. In general, for the sinusoids in Figure 15-2, you have an amplitude V_A, a radian frequency ω, and a phase shift of ϕ given by the following expression:

$$v(t) = V_A \cos(\omega t + \phi)$$
$$v(t) = V_A \,\mathrm{Re}\left\{ e^{j(\omega t + \phi)} \right\} = \mathrm{Re}\left[\underbrace{V_A e^{j\phi}}_{\mathbf{V}}\, e^{j\omega t} \right]$$

Because the radian frequency ω remains the same in a linear circuit, a phasor just needs the amplitude V_A and the phase ϕ to get into polar form:

$$\mathbf{V} = V_A e^{j\phi}$$

To describe a phasor, you need only the amplitude and phase shift (not the radian frequency). Using Euler's formula, the rectangular form of the phasor is

$$\mathbf{V} = V_A \cos\phi + jV_A \sin\phi$$

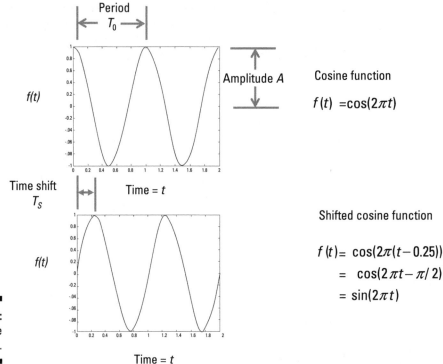

Figure 15-2:
Cosine
functions.

Examining the properties of phasors

One key phasor property is the additive property. If you add sinusoids that
have the *same frequency,* then the resulting phasor is simply the vector sum of
the phasors — just like adding vectors:

$$\mathbf{V} = \mathbf{V}_1 + \mathbf{V}_2 + \cdots \mathbf{V}_N$$

For this equation to work, phasors $\mathbf{V}_1$, $\mathbf{V}_2$, ..., $\mathbf{V}_N$ must have the same frequency. You find this property useful when using Kirchhoff's laws.

Another vital phasor property is the time derivative. The time derivative of a sine wave is another scaled sine wave with the same frequency. Taking the derivative of phasors is an algebraic multiplication of $j\omega$ in the phasor domain. First, you relate the phasor of the original sine wave to the phasor of the derivative:

$$\frac{dv(t)}{dt} = \frac{d}{dt}\left(\mathbf{V}e^{j\omega t}\right)$$
$$= \mathbf{V}\frac{d}{dt}\left(e^{j\omega t}\right)$$

But the derivative of a complex exponential is another exponential multiplied by $j\omega$:

$$\frac{dv(t)}{dt} = \left(j\omega\mathbf{V}\right)e^{j\omega t}$$

Based on the phasor definition, the quantity $(j\omega\mathbf{V})$ is the phasor of the time derivative of a sine wave phasor $\mathbf{V}$. Rewrite the phasor $j\omega\mathbf{V}$ as

$$j\omega\mathbf{V} = \left(\omega e^{j90°}\right)\left(V_A e^{j\theta}\right)$$
$$= \omega V_A e^{j(\theta+90°)}$$

When taking the derivative, you multiply the amplitude V_A by ω and shift the phase angle by 90°, or equivalently, you multiply the original sine wave by $j\omega$. See how the imaginary number j rotates a phasor by 90°?

Working with capacitors and inductors involves derivatives because things change over time. For capacitors, how quickly a capacitor voltage changes directs the capacitor current. For inductors, how quickly an inductor current changes controls the inductor voltage.

Using Impedance to Expand Ohm's Law to Capacitors and Inductors

The concept of impedance is very similar to resistance. You use the concept of impedance to formulate Ohm's law in phasor form so you can apply and extend the law to capacitors and inductors. After describing impedance, you use phasor diagrams to show the phase difference between voltage and current. These diagrams show how the phase relationship between the voltage and current differs for resistors, capacitors, and inductors.

Understanding impedance

For a circuit with only resistors, Ohm's law says that voltage equals current times resistance, or $V = IR$. But when you add storage devices to the circuit, the i-v relationship is a little more, well, complex. Resistors get rid of energy as heat, while capacitors and inductors store energy. Capacitors resist changes in voltage, while inductors resist changes in current. *Impedance* provides a direct relationship between voltage and current for resistors, capacitors, and inductors when you're analyzing circuits with phasor voltages or currents.

Like resistance, you can think of impedance as a proportionality constant that relates the phasor voltage **V** and the phasor current **I** in an electrical device. Put in terms of Ohm's law, you can relate **V, I,** and impedance Z as follows:

$$\mathbf{V} = \mathbf{I}Z$$

The impedance Z is a complex number:

$$Z = R + jX$$

Here's what the real and imaginary parts of Z mean:

- **The real part R is the resistance from the resistors.** You never get back the energy lost when current flows through the resistor. When you have a resistor connected in series with a capacitor, the initial capacitor voltage gradually decreases to 0 if no battery is connected to the circuit. Why? Because the resistor uses up the capacitor's initial stored energy as heat when current flows through the circuit. Similarly, resistors cause the inductor's initial current to gradually decay to 0.

- **The imaginary part X is the *reactance,* which comes from the effects of capacitors or inductors.** Whenever you see an imaginary number for impedance, it deals with storage devices. If the imaginary part of the impedance is negative, then the imaginary piece of the impedance is dominated by capacitors. If it's positive, the impedance is dominated by inductors.

When you have capacitors and inductors, the impedance changes with frequency. This is a big deal! Why? You can design circuits to accept or reject specific ranges of frequencies for various applications. When capacitors or inductors are used in this context, the circuits are called *filters*. You can use these filters for things like setting up fancy Christmas displays with multicolored lights flashing and dancing to the music.

The reciprocal of impedance Z is called the *admittance Y:*

$$Y = \frac{1}{Z} = G + jB$$

The real part G is called the *conductance,* and the imaginary part B is called *susceptance.*

Looking at phasor diagrams

Phasor diagrams explain the differences among resistors, capacitors, and inductors, where the voltage and current are either in phase or out of phase by 90°. A resistor's voltage and current are in phase because an instantaneous change in current corresponds to an instantaneous change in voltage. But for capacitors, voltage doesn't change instantaneously, so even if the current changes instantaneously, the voltage will lag the current. For inductors, current doesn't change instantaneously, so when there's an instantaneous change in voltage, the current lags behind the voltage.

Figure 15-3 shows the phasor diagrams for these three devices. For a resistor, the current and voltage are in phase because the phasor description of a resistor is $V_R = I_R R$. The capacitor voltage lags the current by 90° due to $-j/(\omega C)$, and the inductor voltage leads the current by 90° due to $j\omega L$.

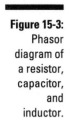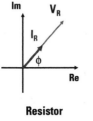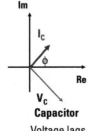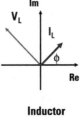

Figure 15-3:
Phasor
diagram of
a resistor,
capacitor,
and
inductor.

| **Resistor** | **Capacitor** | **Inductor** |
| Voltage in phase with current | Voltage lags current by 90° | Voltage leads current by 90° |

Illustration by Wiley, Composition Services Graphics

Putting Ohm's law for capacitors in phasor form

For a capacitor with capacitance C, you have the following current:

$$i = C\frac{dv}{dt}$$

Because the derivative of a phasor simply multiplies the phasor by $j\omega$, the phasor description for a capacitor is

$$\mathbf{I} = j\omega C\mathbf{V} \text{ or } \mathbf{V} = \mathbf{I}\underbrace{\left(\frac{1}{j\omega C}\right)}_{Z_c}$$

The phasor description for a capacitor has a form similar to Ohm's law, showing that a capacitor's impedance is

$$Z_C = \frac{1}{j\omega C} = -j\frac{1}{\omega C}$$

Figure 15-3 shows the phasor diagram of a capacitor. The capacitor voltage lags the current by 90°, as you can see from Euler's formula:

$$-j = e^{-j90°} = \cos(-90°) + j\sin(-90°)$$

Think of the imaginary number j as an operator that rotates a vector by 90° in the counterclockwise direction. A $-j$ rotates a vector in the clockwise direction. You should also note j^2 rotates the phasor by 180° and is equal to -1.

The imaginary component for a capacitor is negative. As the radian frequency ω increases, the capacitor's impedance goes down. Because the frequency for a battery is 0 and a battery has constant voltage, the impedance for a capacitor is infinite. The capacitor acts like an open circuit for a constant voltage source.

Putting Ohm's law for inductors in phasor form

For an inductor with inductance L, the voltage is

$$v = L\frac{di}{dt}$$

The corresponding phasor description for an inductor is

$$\mathbf{V} = \underbrace{j\omega L}_{Z_L}\mathbf{I}$$

The impedance for an inductor is

$$Z_L = j\omega L$$

Figure 15-3 shows the phasor diagram of an inductor. The inductor voltage leads the current by 90° because of Euler's formula:

$$j = e^{j90°} = \cos(90°) + j\sin(90°)$$

The imaginary component is positive for inductors. As the radian frequency ω increases, the inductor's impedance goes up. Because the radian frequency for a battery is 0 and a battery has constant voltage, the impedance is 0. The inductor acts like a short circuit for a constant voltage source.

Tackling Circuits with Phasors

Phasors are great for solving steady-state responses (assuming zero initial conditions and sinusoidal inputs). Under the phasor concept, everything I cover in earlier chapters can be reapplied here. You can take functions of voltages $v(t)$ and currents $i(t)$ described in time to the phasor domain as $\mathbf{V}$ and $\mathbf{I}$. With phasor methods, you can algebraically analyze circuits that have inductors and capacitors, similar to how you analyze resistor-only circuits.

When analyzing circuits in the phasor domain for sine or cosine wave (sinusoidal signal) inputs, use these steps:

1. **Transform the circuit into the phasor domain by putting the sinusoidal inputs and outputs in phasor form.**

2. **Transform the resistors, capacitors, and inductors into their imped-ances in phasor form.**

3. **Use algebraic techniques to do circuit analysis to solve for unknown phasor responses.**

4. **Transform the phasor responses back into their time-domain sinusoids to get the response waveform.**

Using divider techniques in phasor form

In a series circuit with resistors, capacitors, inductors, and a voltage source, you can use phasor techniques to obtain the voltage across any device in the circuit. You can generalize the series circuit and voltage divider concept in Chapter 4 by replacing the resistors, inductors, and capacitors with impedances.

Remember that in a series circuit, you have the same current flowing through each device. When a series circuit is driven by a voltage source, you can find the voltage across each device using voltage divider techniques. This involves multiplying a voltage source by the ratio of the desired device impedance to the total impedance of the series circuit.

The top diagram of Figure 15-4 shows an RLC (resistor, inductor, capacitor) series circuit to illustrate the voltage divider concept and series equivalence:

$$\mathbf{V} = \mathbf{V}_1 + \mathbf{V}_2 + \mathbf{V}_3$$
$$= Z_1\mathbf{I} + Z_2\mathbf{I} + Z_3\mathbf{I}$$
$$= \underbrace{\left(Z_1 + Z_2 + Z_3\right)}_{Z_{EQ}}\mathbf{I}$$

You have an equivalent impedance Z_{EQ} from the three devices:

$$Z_{EQ} = Z_1 + Z_2 + Z_3$$

Here's the equivalent impedance for the RLC series circuit in Figure 15-4:

$$Z_{EQ} = j\omega L + R + \frac{1}{j\omega C} = R + j\left(\omega L - \frac{1}{\omega C}\right)$$

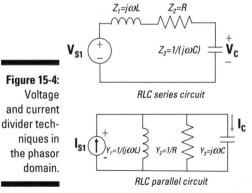

Figure 15-4:
Voltage
and current
divider tech-
niques in
the phasor
domain.

RLC series circuit

RLC parallel circuit

Illustration by Wiley, Composition Services Graphics

To get the voltage $\mathbf{V}_3 = \mathbf{V}_C$, use the voltage divider technique:

$$\mathbf{V}_3 = \frac{Z_3}{Z_{EQ}} \mathbf{V}$$

Now plug in the values for Z_3 and Z_{EQ} to get the capacitor voltage ($\mathbf{V}_3 = \mathbf{V}_C$):

$$\mathbf{V}_3 = \frac{\left(\dfrac{1}{j\omega C}\right)}{R + j\omega L + \left(\dfrac{1}{j\omega C}\right)} \mathbf{V}_{S1}$$

You can also obtain the equivalent impedance for parallel circuits and use the current divider method. (To see how to derive the equivalent resistance and current divider equations, see Chapter 4.) Parallel devices have the same voltage, which helps you get the total admittance (the reciprocal of imped-ance Z):

$$Y_{EQ} = Y_1 + Y_2 + Y_3$$

For the RLC parallel circuit in Figure 15-4, the equivalent admittance is

$$Y_{EQ} = \frac{1}{j\omega L} + \frac{1}{R} + j\omega C$$

To find the capacitor current $\mathbf{I}_3 = \mathbf{I}_C$, use the current divider technique:

$$\mathbf{I}_3 = \mathbf{I}_C = \frac{Y_3}{Y_{EQ}} \mathbf{I}_{S1}$$

Plugging in the values for Y_1 and Y_{EQ}, the capacitor current $\mathbf{I}_3 = \mathbf{I}_C$ is

$$\mathbf{I}_3 = \mathbf{I}_C = \mathbf{I}_{S1} \frac{j\omega C}{\left(\dfrac{1}{j\omega L} + \dfrac{1}{R} + j\omega C \right)}$$

Adding phasor outputs with superposition

Superposition (see Chapter 7) says you can find the phasor output due to one source by turning off other sources; you then get the total output by adding up the individual phasor outputs.

You can use the superposition technique in phasor analysis only if all the independent sources have the same frequency. Superposition doesn't work when you have different frequencies in the independent sources — you treat each source separately to get its steady-state output contribution to the total output.

To see how superposition works with phasors, first look at the top circuit in Figure 15-5. The middle diagram turns off $\mathbf{V}_{S2}$, leaving $\mathbf{V}_{S1}$ as the only voltage source. Use the voltage divider method to get the capacitor voltage $\mathbf{V}_{C1}$ due to $\mathbf{V}_{S1}$:

$$\mathbf{V}_{C1} = \frac{R // \left(\dfrac{1}{j\omega C} \right)}{j\omega L + R // \left(\dfrac{1}{j\omega C} \right)} \mathbf{V}_{S1}$$

where // denotes the parallel connection of capacitor C and resistor R.

The parallel combination of R and C has an equivalent impedance of

$$R // \left(\frac{1}{j\omega C} \right) = \frac{R \left(\dfrac{1}{j\omega C} \right)}{R + \left(\dfrac{1}{j\omega C} \right)}$$

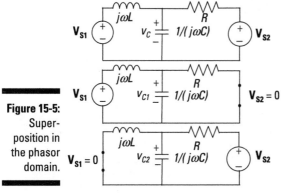

Figure 15-5:
Super-
position in
the phasor
domain.

The bottom diagram of Figure 15-5 turns off $\mathbf{V}_{S1}$, leaving only $\mathbf{V}_{S2}$ turned on.
You use the voltage divider technique with capacitor C and inductor L con-
nected in parallel to get the capacitor voltage due to $\mathbf{V}_{S2}$:

$$\mathbf{V}_{C2} = \frac{j\omega L \, / /\left(\dfrac{1}{j\omega C}\right)}{R + j\omega L \, / /\left(\dfrac{1}{j\omega C}\right)} \mathbf{V}_{S2}$$

The parallel combination of L and C has an equivalence impedance:

$$j\omega L \, / /\left(\frac{1}{j\omega C}\right) = \frac{j\omega L \left(\dfrac{1}{j\omega C}\right)}{j\omega L + \left(\dfrac{1}{j\omega C}\right)}$$

The total output voltage is the sum of $\mathbf{V}_{C1}$ and $\mathbf{V}_{C2}$ due to each source:

$$\mathbf{V}_C = \mathbf{V}_{C1} + \mathbf{V}_{C2}$$

Simplifying phasor analysis with Thévenin and Norton

You can use the Thévenin and Norton equivalents — which I first discuss in Chapter 8 with resistive circuits — in the phasor domain as well. The Thévenin equivalent simplifies a complex array of impedances and independent sources to one voltage source connected in series with one impedance value (a complex number in general). The Norton equivalent simplifies a complex array of impedances and independent sources to one current source connected in parallel with one impedance value. The two equivalents are related by a source transformation. You use the Thévenin and Norton equivalents when you're analyzing different loads to a source circuit.

The Thévenin and Norton equivalents in Figure 15-6 follow the same approach as the one you'd use for resistive circuits. You simply calculate the open-circuit phasor voltage $\mathbf{V}_{OC}$ and short-circuit phasor current $\mathbf{I}_{SC}$ for each equivalent circuit.

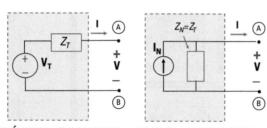

Figure 15-6:
Thévenin
and Norton
equivalents
in phasor
domain.

THÉVENIN EQUIVALENT
Replaces source circuit with one
voltage source and one
impedance.

NORTON EQUIVALENT
Replaces source circuit with one
current source and one
impedance.

Illustration by Wiley, Composition Services Graphics

The following phasor equations are similar to corresponding equations for resistive circuits:

$$\mathbf{V}_{OC} = \mathbf{V}_T = \mathbf{I}_N Z_T$$

$$\mathbf{I}_{SC} = \mathbf{I}_N = \frac{\mathbf{V}_T}{Z_T}$$

Using $\mathbf{V}_{OC}$ and $\mathbf{I}_{SC}$, you find Thévenin impedance Z_T as follows:

$$Z_T = \frac{\mathbf{V}_{OC}}{\mathbf{I}_{SC}}$$

Alternatively, you can calculate the impedance Z_T by looking back to the source circuit between Terminals A and B with all independent sources turned off, as described in Chapter 8.

Figure 15-7 shows a circuit to illustrate the Thévenin equivalent between Terminals A and B. Because you have an open-circuit load, no current flows through resistor R. You can find the open-circuit voltage using the voltage divider technique:

$$\mathbf{V}_{OC} = \frac{\left(\dfrac{1}{j\omega C}\right)}{j\omega L + \left(\dfrac{1}{j\omega C}\right)} \mathbf{V}_{S1}$$

Figure 15-7:
Example
of the
Thévenin
equivalent
in the pha-
sor domain.

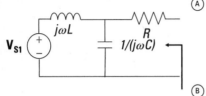

Illustration by Wiley, Composition Services Graphics

Putting a short across Terminals A and B implies that the resistor R and capacitor C are connected in parallel. The current flowing through this combination is

$$\mathbf{I}_1 = \frac{\mathbf{V}_{S1}}{j\omega L + R // \left(\dfrac{1}{j\omega C}\right)}$$

The short-circuit current $\mathbf{I}_{SC}$ flows through $\mathbf{R}$. Using the current divider technique, you get

$$\mathbf{I}_{SC} = \mathbf{I}_1 \left(\frac{\dfrac{1}{j\omega C}}{R + \left(\dfrac{1}{j\omega C}\right)} \right)$$

$$= \frac{\mathbf{V}_{S1}}{j\omega L + R // \left(\dfrac{1}{j\omega C}\right)} \left(\frac{\dfrac{1}{j\omega C}}{R + \left(\dfrac{1}{j\omega C}\right)} \right)$$

You find the impedance Z_T by taking the ratio of $\mathbf{V}_{OC}/\mathbf{I}_{SC}$:

$$Z_T = R + j\omega L \, / / \left(\frac{1}{j\omega C} \right)$$

Getting the nod for nodal analysis

When the circuit is large and complex, node-voltage analysis allows you to reduce the number of equations you need to deal with simultaneously. From the smaller set of node voltages, you can find any voltage or current for any device in the circuit. The node-voltage analysis technique I describe in Chapter 5 also works in the algebraic phasor domain. Figure 15-8 shows an op-amp circuit where you can use node-voltage analysis techniques. (For the scoop on op amps, see Chapter 10.)

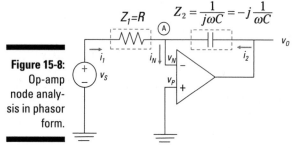

Figure 15-8:
Op-amp node analysis in phasor form.

Illustration by Wiley, Composition Services Graphics

At Node A, you have the following KCL equation:

$$\text{in=out} \quad \rightarrow \quad \frac{\mathbf{V}_S - \overset{=0}{\overbrace{\mathbf{V}_N}}}{Z_1} + \frac{\mathbf{V}_O - \overset{=0}{\overbrace{\mathbf{V}_N}}}{Z_2} = \overset{=0}{\overbrace{\mathbf{I}_N}}$$

For ideal op amps with negative feedback, you have the inverting current $\mathbf{I}_N = 0$ and $\mathbf{V}_N = \mathbf{V}_P = 0$. Solve for the output $\mathbf{V}_O$ in terms of the input $\mathbf{V}_S$:

$$\mathbf{V}_O = -\frac{Z_2}{Z_1} \mathbf{V}_S$$

The output is an inverted input multiplied by the ratio of impedances. If the input impedance Z_1 is due to a resistor and feedback impedance Z_2 is due to a capacitor, then the phasor output V_O is

$$V_O = -\frac{Z_2}{Z_1}V_S = -\frac{\left(\dfrac{1}{j\omega C}\right)}{R}V_S = -\left(\frac{1}{j\omega RC}\right)V_S$$

This equation should look familiar, because it's the integrator of the function waveform $v_S(t)$. You see that the $1/j\omega$ term describes the phasor for an integrator. That's how an integrator is done electronically with op amps — beautiful!

Using mesh-current analysis with phasors

Mesh-current analysis is useful when a circuit has several loops. From the smaller set of mesh currents, you can find any voltage or current for any device in the circuit.

You can open up the mesh analysis approach in Chapter 6 to the phasor domain. You simply replace each device with its phasor impedance and apply KVL for each mesh to develop the mesh current equations. Figure 15-9 helps show the phasor analysis of circuits using mesh current techniques.

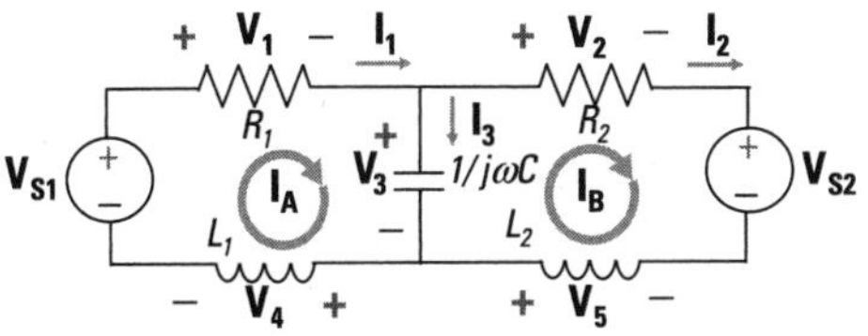

Figure 15-9: Mesh-current analysis using phasors.

Illustration by Wiley, Composition Services Graphics

The circuit has two mesh currents, I_A and I_B, and five devices. For Meshes A and B, KVL produces the following:

Mesh A: $V_1 + V_3 + V_4 = V_{S1}$

Mesh B: $-V_2 + V_3 + V_5 = V_{S2}$

Replace the phasor voltages with the corresponding mesh currents and impedances:

Mesh A: $\mathbf{I}_A\left(R_1 + j\omega L_1\right) + \left(\mathbf{I}_A - \mathbf{I}_B\right)\dfrac{1}{j\omega C} = \mathbf{V}_{S1}$

Mesh B: $-\left(\mathbf{I}_A - \mathbf{I}_B\right)\dfrac{1}{j\omega C} + \mathbf{I}_B\left(R_2 + j\omega L_2\right) = \mathbf{V}_{S2}$

You then collect like terms and rearrange the mesh current equations to put them in standard form:

Mesh A: $\mathbf{I}_A\left(R_1 + j\omega L_1 + \dfrac{1}{j\omega C}\right) - \mathbf{I}_B\dfrac{1}{j\omega C} = \mathbf{V}_{S1}$

Mesh B: $-\mathbf{I}_A\dfrac{1}{j\omega C} + \mathbf{I}_B\left(R_2 + j\omega L_2 + \dfrac{1}{j\omega C}\right) = \mathbf{V}_{S2}$

Convert the equations to matrix form:

$$\begin{bmatrix} R_1 + j\omega L_1 + \dfrac{1}{j\omega C} & -\dfrac{1}{j\omega C} \\[2mm] -\dfrac{1}{j\omega C} & R_2 + j\omega L_2 + \dfrac{1}{j\omega C} \end{bmatrix} \begin{bmatrix} \mathbf{I}_A \\ \mathbf{I}_B \end{bmatrix} = \begin{bmatrix} \mathbf{V}_{S1} \\ \mathbf{V}_{S2} \end{bmatrix}$$

Note the symmetry along the diagonal of the first matrix. For circuits with independent sources, this symmetry is a useful check to verify that your mesh current equations are correct.

You can then use matrix software to solve for the unknown mesh currents $\mathbf{I}_A$ and $\mathbf{I}_B$, which you use to find the device currents and voltages.

Chapter 16

Predicting Circuit Behavior with Laplace Transform Techniques

In This Chapter

▶ Switching domains with the Laplace and inverse Laplace transforms

▶ Defining poles and zeros

▶ Working out a circuit response with Laplace methods

*A*nalyzing the behavior of circuits consisting of resistors, capacitors, and inductors can get complicated because it involves differential equations. Although the classical differential equation approach using calculus is straightforward, the Laplace approach has the advantage of using simpler algebraic techniques. Also, the Laplace transform uncovers properties of circuit behavior you don't normally see using calculus.

In this chapter, I introduce you to the Laplace transform, show you how to find the inverse Laplace transform, and explain how to use the Laplace transform to predict a circuit's behavior.

Getting Acquainted with the Laplace Transform and Key Transform Pairs

The Laplace transform allows you to change a tough differential equation requiring calculus into a simpler problem involving algebra in the *s*-domain (also known as the *Laplace domain*). After finding the transform solution in the *s*-domain, you use the inverse Laplace transform to find the time-domain solution to your original differential equation. In this chapter, finding the inverse Laplace transform basically requires you to look up a transform pair using a table.

In the following equation, the Laplace transform takes a function $f(t)$, described in the time-domain, and transforms it into another function $F(s)$, described in the s-domain.

$$F(s) = \int_{0-}^{\infty} f(t)e^{-st}dt = \mathscr{L}[f(t)]$$

The Laplace transform of $f(t)$, defined as $F(s)$, is a function of the complex frequency variable s, which is defined as

$$s = \sigma + j\omega$$

The preceding equation has a real part σ and an imaginary part ω. The complex variable s is an independent variable in the complex frequency domain, similar to the independent variable t in the time-domain.

Based on the preceding discussion, Figure 16-1 shows the process of applying the Laplace and inverse Laplace transform techniques to solve a problem algebraically.

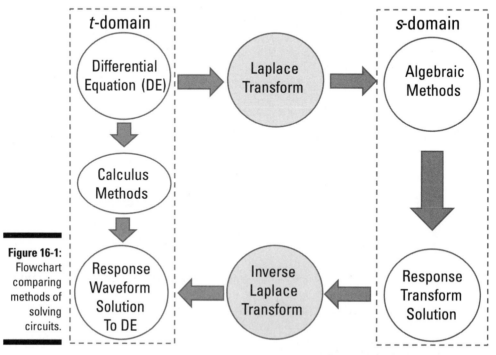

Figure 16-1: Flowchart comparing methods of solving circuits.

Table 16-1 lists the Laplace transform pairs that you'll find most helpful when working with circuits.

Table 16-1	Key Laplace Transform Pairs	
Signal Description	**Time-Domain Waveform, f(t)**	**s-Domain Waveform, F(s)**
Step	$u(t)$	$\dfrac{1}{s}$
Exponential	$\left[e^{-\alpha t}\right]u(t)$	$\dfrac{1}{s+\alpha}$
Impulse	$\delta(t)$	1
Ramp, r(t)	$tu(t)$	$\dfrac{1}{s^2}$
Sine	$\left[\sin\beta t\right]u(t)$	$\dfrac{\beta}{s^2+\beta^2}$
Cosine	$\left[\cos\beta t\right]u(t)$	$\dfrac{s}{s^2+\beta^2}$
Damped Pairs		
Damped ramp	$te^{-\alpha t}u(t)$	$\dfrac{1}{(s+\alpha)^2}$
Damped sine	$\left[e^{-\alpha t}\sin\beta t\right]u(t)$	$\dfrac{\beta}{(s+\alpha)^2+\beta^2}$
Damped cosine	$\left[e^{-\alpha t}\cos\beta t\right]u(t)$	$\dfrac{s+\alpha}{(s+\alpha)^2+\beta^2}$

Here are some key properties you may find helpful when analyzing circuits using the Laplace transform approach:

✔ **Linearity property:**

$$af_1(t)+bf_2(t)\rightarrow a\mathscr{L}\{f_1(t)\}+b\mathscr{L}\{f_2(t)\}=aF_1(s)+bF_2(s)$$

You find the linearity property useful when dealing with partial fraction expansion in later sections.

✔ **Integration property:**

$$\int_0^t f(\tau)d\tau \rightarrow \frac{F(s)}{s}$$

✔ **Differentiation property:**

• First-order:

$$\frac{df(t)}{dt} \rightarrow sF(s) - f(0^-)$$

• Second-order:

$$\frac{d^2f(t)}{dt^2} \rightarrow s^2F(s) - sf(0^-) - f'(0^-)$$

You find the integration and differentiation properties useful when dealing with derivatives and integral relationships of element constraints for capacitors and inductors.

Getting Your Time Back with the Inverse Laplace Transform

Say you're given the transform $F(s)$ in the s-domain. You now need to get back to the time-domain solution $f(t)$, which you get through the inverse Laplace transform of $F(s)$. When you have the simpler transforms, you just find the transform pair that has a form similar to the ones in Table 16-1. When you can't find a transform pair in the table, you need to break up the transform $F(s)$ into simpler transforms using a technique called *partial fraction expansion*. The following sections explain the basic partial fraction expansion method and how to modify the method when you have equations with complex or multiple poles.

Rewriting the transform with partial fraction expansion

When a transform $F(s)$ doesn't match those in Table 16-1, you can use partial fraction expansion to separate it. This method reduces the degree of the denominator of $F(s)$. You find the inverse Laplace transform $f(t)$ by rewriting the ratio of polynomials of $F(s)$ as the sum of simpler fractions, finding the inverse Laplace transform for each fraction, and adding the inverse Laplace transforms together. Here are the basic steps:

1. **Factor the numerator and denominator of $F(s)$.**

   Consider the following transform $F(s)$:

   $$F(s) = 10 \cdot \frac{s^2 + 3s + 2}{s^3 + 17s^2 + 92s + 160}$$

   $$F(s) = \frac{10(s+1)(s+2)}{(s+4)(s+5)(s+8)}$$

   Putting the equation in factored form helps you figure out how to break $F(s)$ into simpler transforms.

2. **Rewrite the factored equation as the sum of fractions, using A, B, C, and so on as placeholders for the numerators.**

   Use each pole factor of $F(s)$ as the denominator of a new fraction. Write A, B, and C as placeholders in the numerators.

   Looking at the poles of the denominator of $F(s)$, you can separate $F(s)$ as follows:

   $$F(s) = \frac{10(s+1)(s+2)}{(s+4)(s+5)(s+8)} = \frac{A}{s+4} + \frac{B}{s+5} + \frac{C}{s+8}$$

   This equation is a partial fraction expansion of $F(s)$.

3. **Find the numerators by solving for the constants.**

   One way to find the constants is to get rid of the denominators. To do so, multiply both sides of the equation by $(s + 4)(s + 5)(s + 8)$:

   $$10(s+1)(s+2) = \left(\frac{A}{s+4} + \frac{B}{s+5} + \frac{C}{s+8}\right)(s+4)(s+5)(s+8)$$

   $$10(s+1)(s+2) = A(s+5)(s+8) + B(s+4)(s+8) + C(s+4)(s+5)$$

   To find A, plug in $s = -4$, which gets rid of the terms that contain B and C:

   $$10(-4+1)(-4+2) = A(-4+5)(-4+8) + 0 + 0$$

   $$10(-3)(-2) = A(1)(4)$$

   $$A = 15$$

   To find B, substitute $s = -5$, which gets rid of the A and C terms:

   $$10(-5+1)(-5+2) = 0 + B(-5+4)(-5+8) + 0$$

   $$10(-4)(-3) = B(-1)(3)$$

   $$B = -40$$

To find C, substitute $s = -8$, which gets rid of the A and B terms:

$$10(-8+1)(-8+2)=0+0+C(-8+4)(-8+5)$$
$$10(-7)(-6)=C(-4)(-3)$$
$$C=35$$

4. **Plug the values of the constants into the partial fraction expansion form of $F(s)$ and find each term's transform pair.**

    Using the values of A, B, and C, you can express the original transform $F(s)$ in the following partial fraction expansion:

    $$F(s)=\frac{15}{s+4}-\frac{-40}{s+5}+\frac{35}{s+8}$$

    In this equation, each of the three simpler terms of the transform follows the mathematical form of an exponential. Using Table 16-1, the terms have the following transform pairs:

    $$\frac{15}{s+4} \quad \leftrightarrow \quad 15e^{-4t}$$

    $$\frac{-40}{s+5} \quad \leftrightarrow \quad -40e^{-5t}$$

    $$\frac{35}{s+8} \quad \leftrightarrow \quad 35e^{-8t}$$

5. **Write the inverse Laplace transform.**

    Based on these pairs, the inverse Laplace transform for $F(s)$ leads to the following transform pair:

    $$F(s)=10\cdot\frac{s^2+3s+2}{s^3+17s^2+92s+160} \quad \leftrightarrow \quad f(t)=15e^{-4t}-40e^{-5t}+35e^{-8t}$$

Expanding Laplace transforms with complex poles

When you have complex poles in the denominator, the Laplace transform function $F(s)$ corresponds to a combination of damped sinusoids. Because $F(s)$ corresponds to damped sinusoids, you can write its partial fraction expansion as follows:

$$F(s)=\frac{As}{(s+\alpha)^2+\beta^2}+\frac{B}{(s+\alpha)^2+\beta^2}$$

You need to determine the constants A and B and then use Table 16-1 to obtain the transform pair. The following steps show how to put the preceding equation in the appropriate form so you can use Table 16-1:

$$F(s) = \frac{As + A\alpha - A\alpha}{(s+\alpha)^2 + \beta^2} + \frac{B}{(s+\alpha)^2 + \beta^2} \qquad \text{(add } A\alpha - A\alpha\text{)}$$

$$F(s) = \frac{As + A\alpha}{(s+\alpha)^2 + \beta^2} + \frac{B - A\alpha}{(s+\alpha)^2 + \beta^2} \qquad \text{(put } -A\alpha \text{ in 2nd term)}$$

$$F(s) = \frac{A(s+\alpha)}{(s+\alpha)^2 + \beta^2} + \frac{[(B-A\alpha)/\beta]\beta}{(s+\alpha)^2 + \beta^2} \qquad \begin{array}{l}\text{(factor out } A \text{ in 1st} \\ \text{term \& multiply by} \\ \beta/\beta \text{ in 2nd term)}\end{array}$$

Now you can use Table 16-1 to get the following transform pair:

$$F(s) = A\frac{(s+\alpha)}{(s+\alpha)^2 + \beta^2} + \left(\frac{B - A\alpha}{\beta}\right)\frac{\beta}{(s+\alpha)^2 + \beta^2} \leftrightarrow$$

$$f(t) = Ae^{-\alpha t}\cos(\beta t) + \left(\frac{B - A\alpha}{\beta}\right)e^{-\alpha t}\sin(\beta t)$$

I know what you're thinking: Enough with the variables already! Your wish is my command. Consider the following transform $F(s)$ and its partial fraction expansion form. (Notice how the numbers are plugged in? You're welcome.)

$$F(s) = \frac{20(s+3)}{(s+1)(s^2+2s+5)} = \frac{As}{s^2+2s+5} + \frac{B}{s^2+2s+5} + \frac{C}{s+1}$$

This equation has a complex pair of poles of $1 + 2j$ and $1 - 2j$ along with a real pole at -1. To match the form found in Table 16-1, you can take the denominators in the first two terms and manipulate them into a perfect square:

$$\frac{20(s+3)}{(s+1)(s^2+2s+5)} = \frac{As}{(s^2+2s+1)+2^2} + \frac{B}{(s^2+2s+1)+2^2} + \frac{C}{s+1}$$

$$\frac{20(s+3)}{(s+1)(s^2+2s+5)} = \frac{As}{(s+1)^2+2^2} + \frac{B}{(s+1)^2+2^2} + \frac{C}{s+1}$$

Clearing out the denominators generates the following equations:

$$20(s+3) = As(s+1) + B(s+1) + C(s^2+2s+5)$$

$$0s^2 + 20s + 60 = (A+C)s^2 + (A+B+2C)s + (B+5C)$$

By equating the coefficients of s^2, s, and the constants on the left and right sides of the preceding equation, you get the following three equations and three unknowns:

$$0 = A + C$$
$$20 = A + B + 2C$$
$$60 = B + 5C$$

Solving for A, B, and C produces the following values: $A = -10$, $B = 10$, and $C = 10$.

You can verify that these values are correct by substituting them into the preceding equations. To apply a transform pair from Table 16-1, substitute the preceding values into the partial fraction expansion form of $F(s)$ to get the following series of algebraic manipulations:

$$F(s) = \frac{-10s}{s^2 + 2s + 5} + \frac{10}{s^2 + 2s + 5} + \frac{10}{s+1}$$

$$= \frac{\overbrace{-10}^{A=-10}\overbrace{(s+1)}^{s+\alpha}}{\underbrace{\left(s^2 + 2s + 1^2\right) + \underbrace{2^2}_{\beta^2}}_{(s+1)^2 = (s+\alpha)^2}} + \frac{\overbrace{\left[(10-(-10)\cdot 1)/2\right]}^{\left|\frac{B-A\alpha}{\beta}\right|=10} \cdot \overbrace{2}^{\beta=2}}{\underbrace{\left(s^2 + 2s + 1^2\right) + \underbrace{2^2}_{\beta^2}}_{(s+1)^2 = (s+\alpha)^2}} + \frac{\overbrace{10}^{C=10}}{s+1}$$

$$= \frac{-10(s+1)}{(s+1)^2 + 2^2} + \frac{10(2)}{(s+1)^2 + 2^2} + \frac{10}{s+1}$$

You can now use Table 16-1 to produce the following inverse Laplace transform of $F(s)$:

$$f(t) = -10e^{-t}\cos(2t) + 10e^{-t}\sin(2t) + 10e^{-t}$$

Dealing with transforms with multiple poles

When you have multiple *poles* — that is, roots in the denominator of $F(s)$ — you need to slightly modify the partial fraction expansion method. With multiple roots, you need to form unique partial fractions with the same poles. To make each fraction unique, you raise the power of the denominator to a specific power. The number of fractions you need with the same poles is equal to the number of poles that have the same value.

You start off with a fraction with the denominator raised to a power of 1. You form another fraction with the denominator raised to the power by incrementing the power (exponent) by 1. You keep forming fractions until you end up with the power that's the same as the number of poles that are equal. So if you have two poles that are the same, then you have one fraction with the polynomial in the denominator raised to a power of 1 and another fraction with the denominator raised to a power of 2.

For example, say you're given the following $F(s)$ with a double pole:

$$F(s) = \frac{8(s+6)}{s(s+4)^2}$$

The double pole is at $s = -4$, and the single pole is at $s = 0$. In this case, the partial fraction expansion for $F(s)$ is

$$F(s) = \frac{8(s+6)}{s(s+4)^2} = \frac{A}{s} + \frac{B}{s+4} + \frac{C}{(s+4)^2}$$

You need to determine the constants A, B, and C. Note that the right side of the equation has a single pole at -4 for the term having B in the numerator and that the third term has a double pole at -4 with C in the numerator. You can easily extend this setup of the partial fraction expansion for more than two poles.

Clearing out the denominators leads to the following expression:

$$8(s+6) = A(s+4)^2 + Bs(s+4) + Cs$$

Substitute $s = 0$ in the preceding equation to find A:

$$8(0+6) = A(0+4)^2 + B(0)(0+4) + C(0)$$
$$48 = 16A + 0 + 0$$
$$A = 3$$

Plug in $s = -4$ to find C:

$$8(-4+6) = A(-4+4)^2 + B(-4)(-4+4) + C(-4)$$
$$16 = 0 + 0 - 4C$$
$$C = -4$$

To find B, you can't use $s = -4$ again. Because you already know $A = 3$ and $C = -4$, you can try any value of s to solve for B. Letting $s = 1$ produces the following expression and value for B:

$$8(1+6) = \underbrace{3}_{A}(1+4)^2 + B(1)(1+4) + \underbrace{(-4)1}_{C}$$

$$56 = 75 + 5B - 4$$

$$B = -3$$

Substituting A, B, and C into $F(s)$ gives you the following expression:

$$F(s) = \frac{8(s+6)}{s(s+4)^2} = \frac{3}{s} + \frac{-3}{s+4} + \frac{-4}{(s+4)^2}$$

Based on Table 16-1, you wind up with the following inverse Laplace transform of $F(s)$:

$$f(t) = 3u(t) - 3e^{-4t} - 4te^{-4t}$$

Understanding Poles and Zeros of F(s)

You can view the Laplace transforms $F(s)$ as ratios of polynomials in the s-domain. If you find the real and complex roots of these polynomials, you can use Table 16-1 to get a general idea of what the waveform $f(t)$ will look like. For example, if the roots are real, then the waveform is exponential. If they're imaginary, then it's a combination of sines and cosines. And if they're complex, then it's a damping sinusoid.

The roots of the polynomial in the numerator of $F(s)$ are *zeros,* and the roots of the polynomial in the denominator are *poles.* The poles result in $F(s)$ blowing up to infinity or being undefined — they're the vertical asymptotes and holes in your graph.

Usually, you create a *pole-zero* diagram by plotting the roots in the s-plane (real and imaginary axes). The pole-zero diagram provides a geometric view and general interpretation of the circuit behavior.

For example, consider the following Laplace transform $F(s)$:

$$F(s) = 10 \cdot \frac{s^2 + 3s + 2}{s^3 + 17s^2 + 92s + 160}$$

This expression is a ratio of two polynomials in s. Factoring the numerator and denominator gives you the following Laplace description $F(s)$:

$$F(s) = 10 \cdot \frac{(s+1)(s+2)}{(s+4)(s+5)(s+8)}$$

The *zeros,* or roots of the numerator, are $s = -1, -2$. The *poles,* or roots of the denominator, are $s = -4, -5, -8$.

Both poles and zeros are collectively called *critical frequencies* because crazy output behavior occurs when $F(s)$ goes to zero or blows up. By combining the poles and zeros, you have the following set of critical frequencies: $\{-1, -2, -4, -5, -8\}$.

Figure 16-2 plots these critical frequencies in the s-plane, providing a geometric view of circuit behavior. In this pole-zero diagram, X denotes poles and O denotes the zeros.

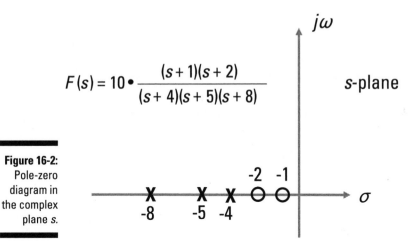

Figure 16-2:
Pole-zero
diagram in
the complex
plane s.

Here are some examples of the poles and zeros of the Laplace transforms, $F(s)$, that you see in Table 16-1. I then follow the examples with pole-zero diagrams — plots of their poles and zeros in the s-plane — in Figure 16-3.

The Laplace transform $F_1(s)$ for a damping exponential has a transform pair as follows:

$$f_1(t) = e^{-\alpha t}u(t) \quad \leftrightarrow \quad F_1(s) = \frac{1}{s+\alpha}$$

The exponential transform $F_1(s)$ has one pole at $s = -\alpha$ and no zeros. Diagram A of Figure 16-3 shows the pole of $F_1(s)$ plotted on the negative real axis in the left half plane.

The sine function has the following Laplace transform pair:

$$f_3(t) = \sin(\beta t) \quad \leftrightarrow \quad F_3(s) = \frac{\beta}{s^2 + \beta^2}$$

The preceding equation has no zeros and two imaginary poles — at $s = +j\beta$ and $s = -j\beta$. Imaginary poles always come in pairs. These two poles are *undamped,* because whenever poles lie on the imaginary axis $j\omega$, the function $f(t)$ will oscillate forever, with nothing to damp it out. Diagram B of Figure 16-3 shows a plot of the pole-zero diagram for a sine function.

A ramp function has the following Laplace transform pair:

$$f_2(t) = tu(t) \quad \leftrightarrow \quad F_2(s) = \frac{1}{s^2}$$

The ramp function has double poles at the origin ($s = 0$) and has no zeros.

Here's a transform pair for a damped cosine signal:

$$f_4(t) = e^{-\alpha t}\cos(\beta t) \quad \leftrightarrow \quad F_4(s) = \frac{s + \alpha}{(s+\alpha)^2 + \beta^2}$$

The preceding equation has two complex poles at $s = \alpha + j\beta$ and $s = \alpha - j\beta$ and one zero at $s = -\alpha$.

Complex poles, like imaginary poles, always come in pairs. Whenever you have a complex pair of poles, the function has oscillations that will be damped out to zero in time — they won't go on forever. The damped sinusoidal behavior consists of a combination of an exponential (due to the real part α of the complex number) and sinusoidal oscillator (due to the imaginary part β of the complex number). Diagram C depicts the pole-zero diagram for a damped cosine.

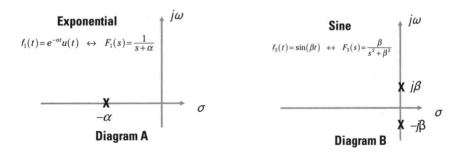

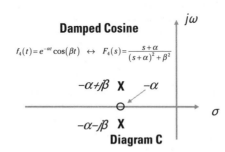

Figure 16-3:
Pole-zero
diagrams
of Laplace
transforms
$F(s)$.

Predicting the Circuit Response with Laplace Methods

Using the Laplace transform as part of your circuit analysis provides you with a different point of view on circuit behavior. One benefit is that the poles of the Laplace transform give you a general idea of the output behavior. Real poles, for instance, indicate exponential output behavior.

All the concepts concerning transient response, frequency response, and the phasor approach developed in Chapters 14 and 15 come together with the Laplace transform. Following are the basic steps for analyzing a circuit using Laplace techniques:

1. **Develop the differential equation in the time-domain using Kirchhoff's laws and element equations.**

2. **Apply the Laplace transformation of the differential equation to put the equation in the s-domain.**

3. **Algebraically solve for the solution, or response transform.**

4. **Apply the inverse Laplace transformation to produce the solution to the original differential equation described in the time-domain.**

To get comfortable with this process, you simply need to practice applying it to different types of circuits. That's why the following sections walk you through each step for three circuits: an RC (resistor-capacitor) circuit, an RL (resistor-inductor) circuit, and an RLC (resistor-inductor-capacitor) circuit.

Working out a first-order RC circuit

Consider the simple first-order RC series circuit in Figure 16-4. To set up the differential equation for this series circuit, you can use Kirchhoff's voltage law (KVL), which says the sum of the voltage rises and drops around a loop is zero. This circuit has the following KVL equation around the loop:

$$-v_s(t)+v_r(t)+v_c(t)=0$$

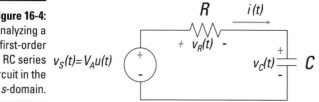

Illustration by Wiley, Composition Services Graphics

Next, formulate the element equation (or *i-v* characteristic) for each device. The element equation for the source is

$$v_S(t)=V_A u(t)$$

Use Ohm's law to describe the voltage across the resistor:

$$v_R(t)=i(t)R$$

The capacitor's element equation is given as

$$i(t)=C\frac{dv_C(t)}{dt}$$

Substituting this expression for $i(t)$ into $v_R(t)$ gives you the following expression:

$$v_R(t)=i(t)R=RC\frac{dv_C(t)}{dt}$$

Substituting $v_R(t)$, $v_C(t)$, and $v_S(t)$ into the KVL equation leads to

$$-v_S(t) + v_R(t) + v_C(t) = 0$$
$$-V_A u(t) + RC\frac{dv_C(t)}{dt} + v_C(t) = 0$$

Now rearrange the equation to get the desired first-order differential equation:

$$RC\frac{dv_C(t)}{dt} + v_C(t) = V_A u(t)$$

Now you're ready to apply the Laplace transformation of the differential equation in the s-domain. The result is

$$\mathscr{L}\left[RC\frac{dv_C(t)}{dt} + v_C(t)\right] = \mathscr{L}\left[V_A u(t)\right]$$
$$\mathscr{L}\left[RC\frac{dv_C(t)}{dt}\right] + \mathscr{L}\left[v_C(t)\right] = \mathscr{L}\left[V_A u(t)\right]$$

On the left, I used the linearity property (from the first section in this chapter) to take the Laplace transform of each term.

For the first term on the left side of the equation, you use the differentiation property (also from the first section), which gives you

$$\mathscr{L}\left[RC\frac{dv_C(t)}{dt}\right] = RC\left[sV_C(s) - V_0\right]$$

This equation uses $V_C(s) = \mathscr{L}\left[v_C(t)\right]$, and V_0 is the initial voltage across the capacitor.

Using Table 16-1, the Laplace transform of a step function provides you with

$$\mathscr{L}\left[V_A u(t)\right] = \frac{V_A}{s}$$

Based on the preceding expressions for the Laplace transforms, the differential equation becomes the following:

$$RC\left[sV_C(s) - V_0\right] + V_C(s) = \frac{V_A}{s}$$

Next, rearrange the equation:

$$\left[s + \frac{1}{RC} \right] V_C(s) = \frac{V_A}{RC} \left(\frac{1}{s} \right) + V_0$$

Solve for the output $V_c(s)$ to get the following transform solution:

$$V_C(s) = \frac{V_A}{RC} \left[\frac{1}{s \left(s + \frac{1}{RC} \right)} \right] + \frac{V_0}{s + \frac{1}{RC}}$$

By performing an inverse Laplace transform of $V_C(s)$ for a given initial condition, this equation leads to the solution $v_C(t)$ of the original first-order differential equation.

On to Step 3 of the process. To get the time-domain solution $v_C(t)$, you need to do a partial fraction expansion for the first term on the right side of the preceding equation:

$$\frac{V_A}{RC} \left[\frac{1}{s \left(s + \frac{1}{RC} \right)} \right] = \frac{A}{s} + \left(\frac{B}{s + \frac{1}{RC}} \right)$$

You need to determine constants A and B. To simplify the preceding equation, multiply both sides by $s(s + 1/RC)$ to get rid of the denominators:

$$\frac{V_A}{RC} = A \left(s + \frac{1}{RC} \right) + Bs$$

Algebraically rearrange the equation by collecting like terms:

$$(A + B)s + \frac{1}{RC}(A - V_A) = 0$$

In order for the left side of the preceding equation to be zero, the coefficients must be zero ($A + B = 0$ and $A - V_A = 0$). For constants A and B, you wind up with $A = V_A$ and $B = -V_A$. Substitute these values into the following equation:

$$\frac{V_A}{RC} \left[\frac{1}{s \left(s + \frac{1}{RC} \right)} \right] = \frac{A}{s} + \left(\frac{B}{s + \frac{1}{RC}} \right)$$

The substitution leads you to:

$$\frac{V_A}{RC}\left(\frac{1}{s\left(s+\dfrac{1}{RC}\right)}\right) = \frac{V_A}{s} + \frac{-V_A}{s+\dfrac{1}{RC}}$$

Now substitute the preceding expression into the $V_C(s)$ equation to get the transform solution:

$$V_C(s) = \frac{V_A}{RC}\left(\frac{1}{s\left(s+\dfrac{1}{RC}\right)}\right) + \frac{V_0}{s+\dfrac{1}{RC}}$$

$$= \frac{V_A}{s} + \frac{-V_A}{s+\dfrac{1}{RC}} + \frac{V_0}{s+\dfrac{1}{RC}}$$

$$= \frac{V_A}{s} + \frac{-V_A}{s+\dfrac{1}{RC}} + \frac{V_0}{s+\dfrac{1}{RC}}$$

That completes the partial fraction expansion. You can then use Table 16-1 to find the inverse Laplace transform for each term on the right side of the preceding equation. The first term has the form of a step function, and the last two terms have the form of an exponential, so the inverse Laplace transform of the preceding equation leads you to the following solution $v_C(t)$ in the time-domain:

$$v_C(t) = V_A u(t) - V_A e^{-\left(\frac{t}{RC}\right)} u(t) + V_0 e^{-\left(\frac{t}{RC}\right)} u(t)$$

$$v_C(t) = V_A\left(1 - e^{-\left(\frac{t}{RC}\right)}\right) u(t) + V_0 e^{-\left(\frac{t}{RC}\right)} u(t)$$

The result shows as time t approaches infinity, the capacitor charges to the value of the input V_A. Also, the initial voltage of the capacitor eventually dies out to zero after a long period of time (about 5 time constants, RC).

Working out a first-order RL circuit

Analyzing an RL circuit using Laplace transforms is similar to analyzing an RC series circuit, which I cover in the preceding section. Figure 16-5 shows you a circuit that has a switch that's been in Position A for a long time. The switch moves to Position B at time $t = 0$.

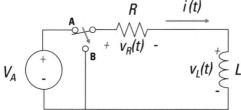

Illustration by Wiley, Composition Services Graphics

For this circuit, you have the following KVL equation:

$$v_R(t) + v_L(t) = 0$$

Next, formulate the element equation (or *i-v* characteristic) for each device. Using Ohm's law to describe the voltage across the resistor, you have the following relationship:

$$v_R(t) = i_L(t)R$$

The inductor's element equation is

$$v_L(t) = L\frac{di_L(t)}{dt}$$

Substituting the element equations, $v_R(t)$ and $v_L(t)$, into the KVL equation gives you the desired first-order differential equation:

$$L\frac{di_L(t)}{dt} + i_L(t)R = 0$$

On to Step 2: Apply the Laplace transform to the differential equation:

$$\mathscr{L}\left[L\frac{di_L(t)}{dt} + i_L(t)R\right] = 0$$

$$\mathscr{L}\left[L\frac{di_L(t)}{dt}\right] + \mathscr{L}\left[i_L(t)R\right] = 0$$

The preceding equation uses the linearity property (see the first section of the chapter), which says you can take the Laplace transform of each term. For the first term on the left side of the equation, you use the differentiation property:

$$\mathscr{L}\left[L\frac{di_L(t)}{dt}\right] = L\left[sI_L(s) - I_0\right]$$

This equation uses $I_L(s) = \mathscr{L}\left[i_L(t)\right]$, and I_0 is the initial current flowing through the inductor.

The Laplace transform of the differential equation becomes

$$I_L(s)R + L\left[sI_L(s) - I_0\right] = 0$$

Solve for $I_L(s)$:

$$I_L(s) = \frac{I_0}{s + \dfrac{R}{L}}$$

For a given initial condition, this equation provides the solution $i_L(t)$ to the original first-order differential equation. You simply perform an inverse Laplace transform of $I_L(s)$ — or look for the appropriate transform pair in Table 16-1 — to get back to the time-domain.

The preceding equation has an exponential form for the Laplace transform pair. You wind up with the following solution:

$$I_L(s) = \frac{I_0}{s + \dfrac{R}{L}} \quad \leftrightarrow \quad i_L(t) = I_0\, e^{-\left(\frac{R}{L}\right)t}$$

The result shows as time t approaches infinity, the initial inductor current eventually dies out to zero after a long period of time — about 5 time constants (L/R).

Working out an RLC circuit

Analyzing an RLC series circuit using the Laplace transform is similar to analyzing an RC series circuit and RL circuit, which I cover in the preceding sections. Figure 16-6 shows you an RLC circuit in which the switch has been open for a long time. The switch is closed at time $t = 0$.

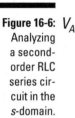

Figure 16-6: Analyzing a second-order RLC series circuit in the s-domain.

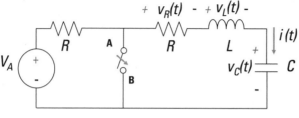

$R = 800\ \Omega$ $C = \dfrac{1}{4.1} \cdot 10^{-5}$ F = 2.439 µF

$L = 1\ \text{H}$ $V_A = 5\ \text{V}$

Illustration by Wiley, Composition Services Graphics

In this circuit, you have the following KVL equation:

$$v_R(t)+v_L(t)+v(t)=0$$

Next, formulate the element equation (or i-v characteristic) for each device. Ohm's law describes the voltage across the resistor (noting that $i(t) = i_L(t)$ because the circuit is connected in series, where $I(s) = I_L(s)$ are the Laplace transforms):

$$v_R(t)=i(t)R$$

The inductor's element equation is given by

$$v_L(t)=L\frac{di_L(t)}{dt}$$

And the capacitor's element equation is

$$v_C(t) = \frac{1}{C} \int_0^t i(\tau) d\tau + v_C(0)$$

Here, $v_C(0) = V_0$ is the initial condition, and it's equal to 5 volts.

Substituting the element equations, $v_R(t)$, $v_C(t)$, and $v_L(t)$, into the KVL equation gives you the following equation (with a fancy name: the *integro-differential equation*):

$$L \frac{di_L(t)}{dt} + i_L(t)R + \frac{1}{C} \int_0^t i(\tau) d\tau + v_C(0) = 0$$

The next step is to apply the Laplace transform to the preceding equation to find an $I(s)$ that satisfies the integro-differential equation for a given set of initial conditions:

$$\mathscr{L}\left[L \frac{di_L(t)}{dt} + i(t)R + \frac{1}{C} \int_0^t i(\tau) d\tau + V_0 \right] = 0$$

$$\mathscr{L}\left[L \frac{di(t)}{dt} \right] + \mathscr{L}\left[i(t)R \right] + \mathscr{L}\left[\frac{1}{C} \int_0^t i(\tau) d\tau + V_0 \right] = 0$$

The preceding equation uses the linearity property (from the first section in this chapter), allowing you to take the Laplace transform of each term.

For the first term on the left side of the equation, you use the differentiation property to get the following transform:

$$\mathscr{L}\left[L \frac{di(t)}{dt} \right] = L\left[sI(s) - I_0 \right]$$

This equation uses $I_L(s) = \mathscr{L}\left[i(t) \right]$, and I_0 is the initial current flowing through the inductor. Because the switch is open for a long time, the initial condition I_0 is equal to zero.

For the second term of the KVL equation dealing with resistor R, the Laplace transform is simply

$$\mathscr{L}\left[i(t)R \right] = I(s)R$$

For the third term in the KVL expression dealing with capacitor C, you have

$$\mathscr{L}\left[\frac{1}{C}\int_0^t i(\tau)d\tau + V_0\right] = \frac{I(s)}{sC} + \frac{V_0}{s}$$

The Laplace transform of the integro-differential equation becomes

$$L\left[sI(s) - I_0\right] + I(s)R + \frac{I(s)}{sC} + \frac{V_0}{s} = 0$$

Rearrange the equation and solve for $I(s)$:

$$I(s) = \frac{sI_0 - \dfrac{V_0}{L}}{s^2 + \dfrac{R}{L}s + \dfrac{1}{LC}}$$

To get the time-domain solution $i(t)$, use Table 16-1 and notice that the preceding equation has the form of a damping sinusoid. Plugging in $I_0 = 0$ and some numbers from Figure 16-6 into the preceding equation gives you

$$I(s) = -\frac{5}{s^2 + 800s + 4.1 \cdot 10^5}$$

$$= -\frac{5}{500}\left[\frac{500}{(s+400)^2 + (500)^2}\right]$$

You wind up with the following solution:

$$i(t) = \left[-0.01e^{-400t}\sin 500t\right]u(t)$$

For this RLC circuit, you have a damping sinusoid. The oscillations will die out after a long period of time. For this example, the time constant is 1/400 and will die out after 5/400 = 1/80 seconds.

Chapter 17

Implementing Laplace Techniques for Circuit Analysis

. . . .

In This Chapter

▶ Starting with basic constraints in the *s*-domain

▶ Looking at voltage and current divider techniques in the *s*-domain

▶ Using superposition, Thévenin, Norton, node voltages, and mesh currents in the *s*-domain

. . . .

*T*his chapter is all about applying Laplace transform techniques in order to study circuits that have voltage and current signals changing with time. That may sound complex, but it's really no more difficult than analyzing resistor-only circuits. You see, the Laplace method converts a circuit to the *s*-domain so you can study the circuit's action using only algebraic techniques (rather than the calculus techniques I show you in Chapters 13 and 14). The algebraic approach in the *s*-domain follows along the same lines as resistor-only circuits, except in place of resistors, you have *s*-domain impedances.

If you need a refresher on impedance or the Laplace transform in general, see Chapters 15 and 16, respectively. Otherwise, I invite you to dive into this chapter, which first has you describe the element and connection constraints in the *s*-domain. You then see how the *s*-domain approach works when you apply voltage and current divider methods, Thévenin and Norton equivalents, node-voltage analysis, and mesh-current analysis.

Starting Easy with Basic Constraints

Connection constraints are those physical laws that cause element voltages and currents to behave in certain ways when the devices are interconnected to form a circuit. You also have constraints on the individual devices themselves, where each device has a mathematical relationship between the voltage across the device and the current through the device. The following sections show you what connection constraints, device constraints, impedances, and admittances wind up looking like in the s-domain.

Connection constraints in the s-domain

Transforming the connection constraints to the s-domain is a piece of cake. Kirchhoff's current law (KCL) says the sum of the incoming and outgoing currents is equal to 0. Here's a typical KCL equation described in the time-domain:

$$i_1(t) + i_2(t) - i_3(t) = 0$$

Because of the linearity property of the Laplace transform (Chapter 16), the KCL equation in the s-domain becomes the following:

$$I_1(s) + I_2(s) - I_3(s) = 0$$

You transform Kirchhoff's voltage law (KVL) in the same way. KVL says the sum of the voltage rises and drops is equal to 0. Here's a classic KVL equation described in the time-domain:

$$v_1(t) + v(t) + v_3(t) = 0$$

Because of linearity, the KVL equation in the s-domain produces

$$V_1(s) + V_2(s) + V_3(s) = 0$$

The basic form of KVL remains the same. Piece of cake!

Device constraints in the s-domain

You can easily transform the *i-v* constraints of devices such as independent and dependent sources, op amps, resistors, capacitors, and inductors to algebraic equations in the *s*-domain. After converting the device constraints, all you need is algebra. I show you how to translate current and voltage relationships to the *s*-domain in the following sections.

Independent and dependent sources

Transforming independent sources is a no-brainer because the *s*-domain has the same form as the time-domain:

$$v_S(t) \rightarrow V_S(s)$$
$$i_S(t) \rightarrow I_S(s)$$

Converting dependent sources is easy, too. Here are the equations for voltage-controlled voltage sources (VCVS), voltage-controlled current sources (VCCS), current-controlled voltage sources (CCVS), and current-controlled current sources (CCCS):

VCVS:	$v_2(t) = \mu v_1(t)$	$\rightarrow$	$V_2(s) = \mu V_1(s)$
VCCS:	$i_2(t) = g v_1(t)$	$\rightarrow$	$I_2(s) = g V_1(s)$
CCVS:	$v_2(t) = r i_1(t)$	$\rightarrow$	$V_2(s) = r i_1(s)$
CCCS:	$i_2(t) = \beta i_1(t)$	$\rightarrow$	$I_2(s) = \beta I_1(s)$

The constants μ, g, r, and β relate the dependent output sources $V_2(s)$ and $I_2(s)$ controlled by input variables $V_1(s)$ and $I_1(s)$. (For more information on dependent sources, see Chapter 9.)

Passive elements: Resistors, capacitors, and inductors

For resistors, capacitors, and inductors, you convert their *i-v* relationships to the *s*-domain using Laplace transform properties, such as the integration and derivative properties (which you find in Chapter 16):

Resistor:	$v_R(t) = R i_R(t)$	$\rightarrow$	$V_R(s) = R I_R(s)$
Capacitor:	$v_C(t) = \int_0^t i_C(\tau)\, d\tau$	$\rightarrow$	$V_C(s) = \dfrac{1}{sC} I_C(s) + \dfrac{v_C(0)}{s}$
Inductor:	$v_2(t) = L\dfrac{d i_L(t)}{dt}$	$\rightarrow$	$V_L(s) = sLI_L(s) - Li_L(0)$

The preceding three equations on the right are *s*-domain models that use voltage sources for the initial capacitor voltage $v_C(0)$ and initial inductor current $i_L(0)$.

You can rewrite these equations in the s-domain to model the initial conditions, $v_C(0)$ and $i_L(0)$, as current sources:

$$\text{Resistor:} \quad V_R(s) = RI_R(s) \rightarrow I_R(s) = \left(\frac{1}{R}\right)V_R(s)$$

$$\text{Capacitor:} \quad V_C(s) = \frac{1}{sC}I_C(s) + \frac{v_C(0)}{s} \rightarrow I_C(s) = (sC)V_C(s) - Cv_C(0)$$

$$\text{Inductor:} \quad V_L(s) = sLI_L(s) - Li_L(0) \rightarrow I_L(s) = \left(\frac{1}{sL}\right)V_L(s) + \frac{i_L(0)}{s}$$

You see there are no integrals or derivatives in the s-domain.

The middle column of Figure 17-1 shows the constraints of the passive devices in the time-domain being converted to the s-domain. The left column shows initial conditions modeled as voltage sources in the s-domain, and the right column shows initial conditions modeled as current sources in the s-domain.

Taking the initial conditions into account in the s-domain analysis for capacitors and inductors is a big deal because it expedites the analysis. When you transform differential equations into the s-domain, you deal with input sources and initial conditions simultaneously.

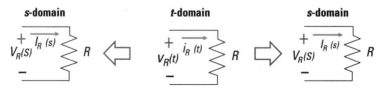

s-domain	t-domain	s-domain
Series model for capacitor and inductor Initial conditions modeled as voltage sources		**Parallel model for capacitor and inductor** Initial conditions modeled as current sources

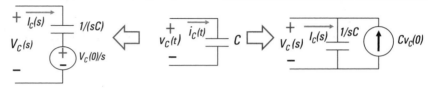

Figure 17-1: The s-domain models of passive devices.

Illustration by Wiley, Composition Services Graphics

Op-amp devices

The constraints of ideal operational amplifiers are unchanged in form in the *s*-domain:

Voltage constraint: $v_p(t) = v_N(t) \rightarrow V_P(s) = V_N(s)$

Current constraint: $i_p(t) = i_N(t) = 0 \rightarrow I_p(s) = I_N(s) = 0$

Impedance and admittance

Impedance Z (see Chapter 15) relates the voltage and current described in the *s*-domain when initial conditions are set to 0. The following algebraic form of the *i-v* relationship describes impedance in the *s*-domain:

$$V(s) = Z(s)I(s)$$

Admittance Y is the reciprocal of the impedance; it's useful when you're analyzing parallel circuits:

$$Y(s) = \frac{1}{Z(s)}$$

In the *s*-domain for zero initial conditions, the element constraints, impedances $Z(s)$, and admittances $Y(s)$ for the passive devices are as follows:

Resistor: $V_R(s) = RI_R(s) \rightarrow Z_R(s) = R$ or $Y_R(s) = \dfrac{1}{R}$

Capacitor: $V_C(s) = \dfrac{1}{sC} I_C(s) \rightarrow Z_C(s) = \dfrac{1}{sC}$ or $Y_C(s) = sC$

Inductor: $V_L(s) = sLI_L(s) \rightarrow Z_L(s) = sL$ or $Y_L(s) = \dfrac{1}{sL}$

Now you're ready to start analyzing circuits in the *s*-domain — without having to rely on calculus.

Seeing How Basic Circuit Analysis Works in the s-Domain

Circuit analysis techniques in the s-domain are powerful because you can treat a circuit that has voltage and current signals changing with time as though it were a resistor-only circuit. That means you can analyze the circuit algebraically, without having to mess with integrals and derivatives. In the following sections, you see how to apply voltage and current divider methods in the s-domain.

Applying voltage division with series circuits

You can put voltage divider techniques to work when dealing with series circuits, as Chapter 4 explains. To use voltage division in the s-domain, you simply replace the resistors with the impedances of devices connected in series. The following voltage divider equation is for three passive devices in a series circuit:

$$v_1(t) = v_s(t)\left(\frac{R_1}{R_1 + R_2 + R_3}\right) \to V_1(s) = V_S(s)\left(\frac{Z_1(s)}{Z_1(s) + Z_2(s) + Z_3(s)}\right)$$

The output voltage $V_1(s)$ is based on the voltage source $V_S(s)$ and on the ratio of the desired impedance $Z_1(s)$ to the total impedance.

Figure 17-2 illustrates the voltage divider for a series circuit for zero initial conditions: $i_L(0) = 0$ and $v_C(0) = 0$. You can find the output transform of the capacitor voltage using the voltage divider equation:

$$V_C(s) = V_S(s)\left(\frac{\frac{1}{sC}}{R + sL + \frac{1}{sC}}\right)$$

In a similar way, the voltage transform across the inductor is

$$V_L(s) = V_S(s)\left(\frac{sL}{R + sL + \frac{1}{sC}}\right)$$

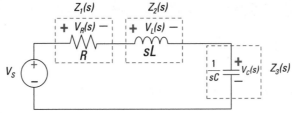

Figure 17-2:
A series
circuit and
the voltage
divider tech-
nique in the
s-domain.

Illustration by Wiley, Composition Services Graphics

And the voltage transform across the resistor is

$$V_R(s) = V_S(s) \left(\frac{R}{R + sL + \dfrac{1}{sC}} \right)$$

That's all there is to it. You may need to do more algebraic gymnastics to simplify other circuits, but you still don't need calculus. To get back to a time-domain description, you need to do a partial fraction expansion; then you look up the inverse Laplace transforms in the table in Chapter 16.

In many cases, you just want to predict what the output is when you're given a particular input. When you know the *transfer function,* which is the ratio between the output transform and the input transform, you can multiply the transfer function by the input voltage to find the output. As a result, you can rewrite the transform of the capacitor voltage as a ratio of polynomials:

$$V_C(s) = V_S(s) \left(\frac{1}{LCs^2 + RCs + 1} \right) = V_S(s) \left(\frac{\dfrac{1}{LC}}{s^2 + \dfrac{R}{L}s + \dfrac{1}{LC}} \right)$$

The denominator is simply a quadratic equation, and the roots of the equation shape the circuit behavior.

Similarly, you can rewrite the transform of the resistor and inductor voltages as a ratio of polynomials.

Turning to current division for parallel circuits

To use current division for parallel circuits having passive devices, all you have to do in the s-domain is replace the conductances with admittances. The following current divider equation is for three passive devices connected in parallel:

$$i_1(t) = i_s(t)\left(\frac{G_1}{G_1 + G_2 + G_3}\right) \rightarrow I_1(s) = I_S(s)\left(\frac{Y_1(s)}{Y_1(s) + Y_2(s) + Y_3(s)}\right)$$

The output current $I_1(s)$ is based on the current source $I_S(s)$ and the ratio of the desired admittance $Y_1(s)$ to the total admittance.

Figure 17-3 illustrates the current divider technique for a parallel circuit for zero initial conditions: $i_L(0) = 0$ and $v_C(0) = 0$. You can find the output transform of the inductor current using the current divider equation:

$$I_L(s) = I_S(s)\left(\frac{\dfrac{1}{sL}}{G + sC + \dfrac{1}{sL}}\right)$$

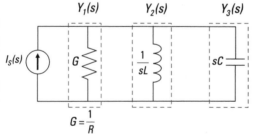

Figure 17-3: Parallel circuits and the current divider technique in the s-domain.

Illustration by Wiley, Composition Services Graphics

In the same way, you get the transform of the capacitor and conductance (or resistor) currents using the current divider technique:

$$I_C(s) = I_S(s)\left(\frac{sC}{G + sC + \dfrac{1}{sL}}\right)$$

$$I_R(s) = I_S(s)\left(\frac{G}{G + sC + \dfrac{1}{sL}}\right)$$

Note that the results resemble the form for series circuits using voltage divider techniques. Neat and simple in the *s*-domain — thank you, Pierre Laplace!

Conducting Complex Circuit Analysis in the s-Domain

In the time-domain, analyzing circuits with resistors, inductors, and capacitors involves integrals and derivatives. You use a simpler algebraic approach by describing and analyzing such circuits in the *s*-domain, as I show you next. The following sections cover node-voltage analysis, mesh-current analysis, superposition, and Norton and Thévenin equivalents in the *s*-domain.

Using node-voltage analysis

In the *s*-domain, node-voltage analysis works the same way as it does for resistor-only circuits, but this time you replace a device with its impedance. Node-voltage analysis (see Chapter 5) allows you to work with a smaller set of equations and unknowns that you need to deal with simultaneously. The unknown variables are called *node voltages*. After you find the unknown voltages, you can find the voltages and currents for each device.

Look at Figure 17-4, which shows a circuit at zero state using an op amp, resistor, and capacitor. You need to find the transfer function $V_O(s)/V_S(s)$. Applying KCL at Node A produces the following:

$$\frac{V_N - V_S}{R_1} + \frac{V_N - V_O}{R_2 + \frac{1}{sC}} + I_N = 0$$

Figure 17-4:
Op-amp node-voltage analysis in the *s*-domain.

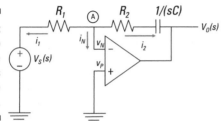

Illustration by Wiley, Composition Services Graphics

For an ideal op amp, $I_N = 0$ and $V_N = V_P = 0$ because V_P is connected to ground. The KCL equation becomes

$$\frac{V_S}{R_1} = -\frac{V_O}{R_2 + \dfrac{1}{sC}}$$

After some algebra, you have the transfer function $V_O(s)/V_S(s)$:

$$\frac{V_O(s)}{V_S(s)} = -\frac{R_2 + \dfrac{1}{sC}}{R_1} = -\left(\frac{R_2}{R_1}\right)\left(\frac{s + \dfrac{1}{R_2C}}{s}\right)$$

If the input is a step input $u(t)$ and its transform is $V_S(s) = 1/s$, the output transform becomes

$$V_O(s) = -\left(\frac{R_2}{R_1}\right)\left(\frac{s + \dfrac{1}{R_2C}}{s^2}\right) = -\left(\frac{R_2}{R_1}\right)\left(\frac{1}{s} + \frac{1}{R_2Cs^2}\right)$$

Use the table in Chapter 16 to get the inverse Laplace transform:

$$v_O(t) = -\left(\frac{R_2}{R_1} + \frac{1}{R_1C}r(t)\right)u(t)$$

The output $v_O(t)$ is a combination of a ramp and a step input. You get this when the circuit acts like an inverting amplifier and an integrator. You have a ramp resulting from the integration of a step input. The inverting amplifier comes into play when the capacitor acts like a short circuit, which occurs at high frequencies for sinusoidal inputs.

Using mesh-current analysis

Mesh-current analysis (see Chapter 6), which is useful when a circuit has several loops, works the same way in the s-domain as it does in resistor-only circuits. You simply use a device's impedance to work the problem. After you solve for the mesh currents, you can find the voltage and current for each device.

Consider the circuit in Figure 17-5. You want to formulate the mesh current equation and solve for the zero-input and zero-state responses. The circuit is transformed into the *s*-domain.

For mesh-current analysis, you need to use the voltage source model of initial conditions; this will give you a circuit with voltage sources. You have the following mesh equations for Loops A and B:

Mesh A: $\left(R_1 + \dfrac{1}{sC} \right) I_A - \left(\dfrac{1}{sC} \right) I_B = V_S(s)$

Mesh B: $-\left(\dfrac{1}{sC} \right) I_A + \left(R_2 + sL + \dfrac{1}{sC} \right) I_B = 0$

Figure 17-5:
Mesh-
current
analysis
in the
s-domain.

Illustration by Wiley, Composition Services Graphics

The matrix software should give you the following result, the algebraic equivalent for $I_B(s)$:

$$I_B(s) = \dfrac{V_S(s)}{sC\left[\left(R_1 + \dfrac{1}{sC} \right)\left(R_2 + sL + \dfrac{1}{sC} \right) - \left(\dfrac{1}{sC} \right)^2 \right]}$$

Using superposition and proportionality

The superposition concept basically says you can take an output *v* as a combination of weighted inputs. When applied to resistor circuits, the superposition concept (presented in Chapter 7) is described as

$$v_o(t) = K_1 v_1 + K_2 v_2 + \cdots + K_n v_n$$

You can apply the same concept to linear circuits in the *s*-domain just by replacing the weighted constants with rational functions of *s*. Then you can look for the response as a sum of the zero-input response due to initial conditions with inputs turned off and the zero-state response due to external sources (inputs) with initial conditions turned off, which means no energy is stored. (You can review these two concepts in Chapters 13 and 14.) You turn off voltage sources by replacing them with short circuits and turn off current sources by replacing them with open circuits.

To see how to use superposition in the *s*-domain, check out Figure 17-6 where $v_s(t)$ is a step input $u(t)$. The upper-left diagram describes an RC series circuit in the time domain, and the bottom-left diagram shows the same circuit described in the *s*-domain. I use this example to kill two birds with one stone: I apply superposition to find the zero-state $V_{ZS}(s)$ or $I_{ZS}(s)$ and the zero-input $V_{ZI}(s)$ or $I_{ZI}(s)$ transform responses, and I show you how to solve the problem by converting a differential equation or integral equation into the Laplace transform.

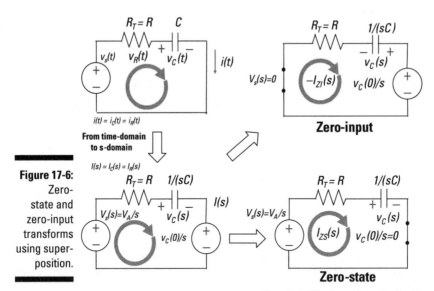

Figure 17-6: Zero-state and zero-input transforms using superposition.

Illustration by Wiley, Composition Services Graphics

First, you need to turn off the input source by replacing the voltage source with a short circuit, as the top-right diagram in Figure 17-6 shows. The result is the zero-input response:

$$I_{zi}(s) = -\frac{\dfrac{v_c(0)}{s}}{R + \dfrac{1}{sC}} = -\frac{\dfrac{v_c(0)}{R}}{s + \dfrac{1}{RC}}$$

The minus sign appears because the current is opposite to the assigned current direction in Figure 17-6. The pole at $s = -1/(RC)$ comes from the circuit. Next, you need to turn off the initial condition modeled as a voltage source by replacing it with a short circuit. You see the zero-state diagram in the lower right of Figure 17-6. You now have the zero-state response for a step input:

$$I_{zs}(s) = \frac{\frac{V_A}{s}}{R + \frac{1}{sC}} = \frac{\frac{V_A}{R}}{s + \frac{1}{RC}}$$

The pole for the zero-state response is $s = -1/(RC)$ from the circuit. Now use superposition to get the total response. Superposition says the total response is the sum of the zero-state and zero-input outputs:

$$I(s) = I_{zi}(s) + I_{zs}(s)$$
$$= -\frac{\frac{v_C(0)}{R}}{s + \frac{1}{RC}} + \frac{\frac{V_A}{R}}{s + \frac{1}{RC}}$$

The table in Chapter 16 tells you that the inverse Laplace transform is an exponential. The inverse Laplace transform of $I(s)$ gives you the time response $i(t)$:

$$i(t) = -\frac{v_C(0)}{R}e^{-t/RC}u(t) + \frac{V_A}{R}e^{-t/RC}u(t)$$

This Laplace stuff really works! Calculus doesn't come into play at all — all you need to do is look up transform pairs in a table to get the time response.

Now take a look at the lower-right circuit in Figure 17-6, which describes the circuit in zero-state in the s-domain. The differential equation for this circuit is based on KVL, given the capacitor voltage as an output variable and replacing forcing function $v_T(t)$ with a step input $V_A u(t)$:

$$v_T(t) = RC\frac{dv(t)}{dt} + v(t)$$
$$V_A u(t) = RC\frac{dv_C(t)}{dt} + v_C(t)$$

Taking the Laplace transform of this equation gives you

$$\frac{V_A}{s} = RC\left(sV_C(s) - v_C(0)\right) + V_C(s)$$

Solve for the transform of the capacitor voltage $V_C(s)$:

$$V_C(s)(RCs + 1) = \frac{V_A}{s} + RCv_C(0)$$

$$V_C(s) = \left[\frac{1}{s(RCs + 1)}\right]V_A + \left[\frac{RC}{RCs + 1}\right]v_C(0)$$

The preceding equation shows you how the forcing function $V_A u(t)$ and the initial condition $v_C(0)$ are taken into account with one step based on the s-domain techniques. Performing a partial fraction expansion on the preceding equation gives you

$$V_C(s) = \frac{V_A}{s} - \frac{V_A}{s + \dfrac{1}{RC}} + \frac{v_C(0)}{s + \dfrac{1}{RC}}$$

Now take the inverse Laplace transform using the table in Chapter 16 to get the capacitor voltage response $v_C(t)$ in the time-domain:

$$v_C(t) = V_A\left(1 - e^{-\left(\frac{t}{RC}\right)}\right) + v_C(0)e^{-\left(\frac{t}{RC}\right)}$$

Taking the derivative of $v_C(t)$ leads you to the capacitor current:

$$i_C(t) = C\frac{dv_C(t)}{dt}$$

$$= \frac{V_A}{R}e^{-\left(\frac{t}{RC}\right)} - \frac{v_C(0)}{R}e^{-\left(\frac{t}{RC}\right)}$$

You get the same capacitor current $i_C(t)$, whether you transform the circuit or transform the differential equation. If you don't like to take the derivative, you can start describing the circuit as an integral equation that just involves the capacitor current $i_C(t)$.

If you use the capacitor current $i_C(t)$ as the output variable, then the KVL equation becomes

$$V_A u(t) = Ri_C(t) + \frac{1}{C}\int i_C(t)dt + v_C(0)$$

Next, perform a Laplace transformation of the preceding equation:

$$\frac{V_A}{s} = RI_C(s) + \frac{I_C(s)}{sC} + \frac{v_C(0)}{s}$$

Solve for the capacitor current $I_C(s)$:

$$\left(R + \frac{1}{sC}\right)I_C(s) = \frac{V_A(s)}{s} - \frac{v_C(0)}{s}$$

$$I_C(s) = \frac{V_A(s)}{R\left(s + \frac{1}{RC}\right)} - \frac{v_C(0)}{R\left(s + \frac{1}{RC}\right)}$$

Again, see how the forcing transform $V_A(s)$ and initial condition $v_C(0)$ are neatly separated components for the capacitor current $I_C(s)$.

Finally, take the inverse Laplace transform of the preceding equation:

$$i_C(t) = \frac{V_A}{R}e^{-\left(\frac{t}{RC}\right)} - \frac{v_C(0)}{R}e^{-\left(\frac{t}{RC}\right)}$$

Using the Thévenin and Norton equivalents

The Thévenin equivalent I present in Chapter 8 simplifies a circuit to one voltage source $v_T(t)$ and one single resistor R_T. Extending the concept to circuits described in the s-domain means replacing the Thévenin resistance R_T with an impedance $Z_T(s)$.

Similarly, the Norton equivalent replaces a complex circuit with a single current source $i_N(t)$ in parallel with the Norton resistor $R_N = R_T$. Extending the Norton concept to the s-domain means replacing the Norton resistance R_N with the impedance $Z_N(s) = Z_T(s)$. Figure 17-7 gives you the visual of how the Thévenin and Norton equivalents reduce circuits in the s-domain.

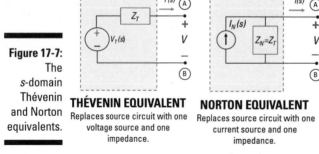

Figure 17-7: The *s*-domain Thévenin and Norton equivalents.

THÉVENIN EQUIVALENT
Replaces source circuit with one voltage source and one impedance.

NORTON EQUIVALENT
Replaces source circuit with one current source and one impedance.

Take a look at the circuit in its zero state in Figure 17-8. The Thévenin and Norton equivalents are related by a source transformation, so use a source transformation to the left of Points A and B. The source transformation converts the Norton source circuit consisting of the independent current source $I_N = I_1(s)$ in parallel with an impedance $Z_N = R$ to a Thévenin equivalent. The Thévenin equivalent consists of a voltage source $V_T = I_N Z_N = RI_1(s)$ in series with $Z_T = Z_N = R$.

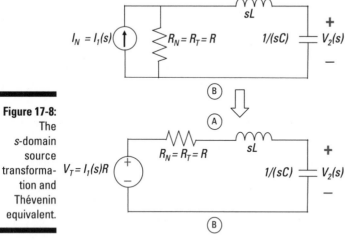

Figure 17-8: The *s*-domain source transformation and Thévenin equivalent.

You use voltage division to find the relationship between the output $V_2(s)$ and the input $I_1(s)$ in the s-domain:

$$V_2(s) = \left[\frac{\frac{1}{sC}}{R + sL + \frac{1}{sC}} \right] \underbrace{RI_1(s)}_{V_T(s)}$$

$$= \left[\frac{R}{LCs^2 + RCs + 1} \right] I_1(s)$$

Factoring out the coefficient LC in the denominator gives you

$$V_2(s) = \left[\frac{R/LC}{s^2 + \frac{R}{L}s + \frac{1}{LC}} \right] I_1(s)$$

To emphasize the Thévenin equivalent when you have circuits with capacitors and inductors, take a look at the bottom diagram of Figure 17-8. The equivalent Thévenin impedance looking to the left from the capacitor terminals is simply the series connection of resistor R and the inductor impedance sL (or mathematically, $Z_T = R + sL$).

Chapter 18

Focusing on the Frequency Responses

. . . .

In This Chapter

▶ Understanding frequency response and types of filters

▶ Interpreting Bode plots

▶ Using circuits to create high-pass, low-pass, band-pass, and band-reject filters

. . . .

*W*hen you hear your favorite music coming from various instruments and melodic voices, the unique sounds you hear consist of many frequencies. In a stereo system, you can adjust the low-frequency and high-frequency sounds by adjusting a stereo equalizer. Equalizers adjust the volume of a specific band of frequencies relative to others. They're often used to boost the bass guitar on bass-hungry speakers or to bring out the vocals of a favorite singer.

With a combination of resistors, capacitors, and inductors, you can select or reject a range of frequencies. As a result, you can pick out frequencies to boost or cut. For audio applications, you can adjust the bass, treble, or mid-range frequencies to get the sound quality you like best. You also find wide applications of frequency response and filtering in communication, control, and instrumentation systems.

How is this all possible? A major component found in older entertainment systems is an electronic filter that shapes the frequency content of signals. You can describe low-pass filters, high-pass filters, band-pass filters, and band-reject filters based on simple circuits. This serves as a foundation for more-complex filters to meet more stringent requirements.

What happens when you want to study a range of frequencies? You use Bode plots. Bode plots help you visualize how poles and zeros affect the frequency response of a circuit.

This chapter shows you what different filters do, explains how Bode plots work, and shows how you can create filters by connecting resistors, inductors, and capacitors.

Describing the Frequency Response and Classy Filters

You find the sinusoidal steady-state output of the filter by evaluating the transfer function $T(s)$ at $s = j\omega$. The transfer function relates the input and output signals in the s-domain and assumes zero initial conditions. The radian frequency ω is a variable that stands for the frequency of the sinusoidal input. After you substitute the $s = j\omega$ into $T(s)$, the transfer function becomes a ratio of complex numbers $T(j\omega)$.

Because the function $T(j\omega)$ is a complex number for all frequencies, you can determine the gain $|T(j\omega)|$ and phase $\theta(j\omega)$. Here are the gain and phase relationships:

$$|T(j\omega)| = \frac{\text{Output amplitude}}{\text{Input amplitude}}$$

$$\theta(j\omega) = \angle T(j\omega)$$
$$= \text{Output phase} - \text{Input phase}$$

You can present the gain and phase as a function of frequency ω graphically, as in Figure 18-1. This figure shows an approximation of a typical filter. In a *passband* region, the gain function has nearly constant gain for a range of frequencies. In the *stopband* region, the gain is significantly reduced for a range of frequencies.

Figure 18-1: Gain and phase plots of the frequency response.

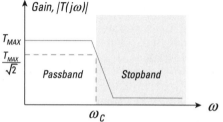

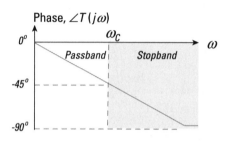

Illustration by Wiley, Composition Services Graphics

For nonideal filters, a *transition* region occurs between adjacent passband and stopband regions. The cutoff frequency ω_C occurs within the transition region, according to a prescribed definition. One widely used definition says the *cutoff* frequency occurs when the passband gain is decreased by a factor of 0.707 from a maximum value T_{MAX}. The mathematical condition for ω_C is therefore

$$\left|T(j\omega)\right| = \frac{1}{\sqrt{2}} T_{MAX} = 0.707 \cdot T_{MAX}$$

At the cutoff frequency, the output power has dropped to one half of its maximum passband value. Here, the passband includes those frequencies where the relative power is greater than the half-power point (0.707 of the maximum value of the transfer function). Frequencies that are less than the half-power point fall in the stopband.

The following sections introduce you to four types of filters. The filters differ in whether they block the frequencies above or below the cutoff frequencies or allow them to pass.

Low-pass filter

The low-pass filter has a gain response with a frequency range from zero frequency (DC) to ω_C. Any input that has a frequency below the cutoff frequency ω_C gets a pass, and anything above it gets attenuated or rejected. The gain approaches zero as frequency increases to infinity.

Figure 18-2 shows the frequency response of a low-pass filter. The input signal has equal amplitudes at frequencies ω_1 and ω_2. After passing through the low-pass filter, the output amplitude at ω_1 is unaffected because it's below the cutoff frequency ω_C. However, at ω_2, the signal amplitude is significantly decreased because it's above ω_C.

Figure 18-2:
Gain response of a low-pass filter.

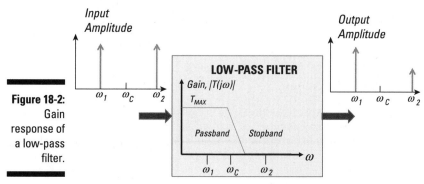

Illustration by Wiley, Composition Services Graphics

High-pass filter

The high-pass filter has a gain response with a frequency range from ω_C to infinity. Any input having a frequency below the cutoff frequency ω_C gets attenuated or rejected. Anything above ω_C passes through unaffected.

Figure 18-3 shows the frequency response of a high-pass filter. The input signal has equal amplitude at frequencies ω_1 and ω_2. After passing through the high-pass filter, the output amplitude at ω_1 is significantly decreased because it's below ω_C, and at ω_2, the signal amplitude passes through unaffected because it's above ω_C.

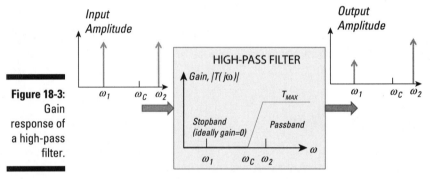

Figure 18-3: Gain response of a high-pass filter.

Illustration by Wiley, Composition Services Graphics

Band-pass filters

The *band-pass filter* has a gain response with a frequency range from ω_{C1} to ω_{C2}. Any input that has frequencies between ω_{C1} and ω_{C2} gets a pass, and anything outside this range gets attenuated or rejected.

Figure 18-4 shows the frequency response of a band-pass filter. The input signal has equal amplitude at frequencies ω_1, ω_2, and ω_3. After passing through the band-pass filter, the output amplitudes at ω_1 and ω_3 are significantly decreased because they fall outside the desired frequency range, while the frequency at ω_2 is within the desired range, so its signal amplitude passes through unaffected.

You can think of the band-pass filter as a series or cascaded connection of a low-pass filter with frequency ω_{C2} and a high-pass filter with frequency ω_{C1}. The bottom diagram of Figure 18-4 shows how the cascade connection of a low-pass filter and high-pass filter forms a band-pass filter. Although the figure shows the low-pass filter before the high-pass filter, the order of the filters doesn't matter.

If you're going to do a quick-and-dirty design of a band-pass filter based on a low-pass filter and high-pass filter, make sure you select the right cutoff frequencies. In Figure 18-4, if you give the low-pass filter a lower cutoff frequency of ω_{C1} and the high-pass filter an upper cutoff frequency of ω_{C2}, you'll get a very small signal at the output. What you'll design in that case is a *no-pass* filter — everything gets rejected.

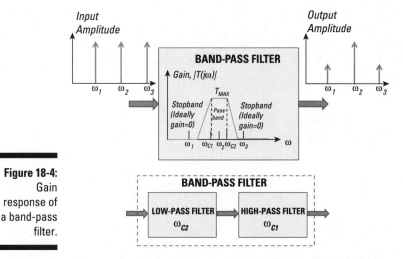

Figure 18-4: Gain response of a band-pass filter.

Illustration by Wiley, Composition Services Graphics

Band-reject filters

The *band-reject filter,* or *bandstop filter,* has a gain response with a frequency range from zero to ω_{C1} and from ω_{C2} to infinity. Any input that has frequencies between ω_{C1} and ω_{C2} gets significantly attenuated, and anything outside this range gets a pass.

Figure 18-5 shows the frequency response of a band-reject filter. The input signal has equal amplitude at frequencies ω_1, ω_2, and ω_3. After passing through the band-reject filter, the output amplitude at ω_1 and ω_3 is unaffected because those frequencies fall outside the range of ω_{C1} to ω_{C2}. But at ω_2, the signal amplitude gets attenuated because it falls within this range.

You can think of the band-pass filter as a parallel connection of a low-pass filter with cutoff frequency ω_{C1} and a high-pass filter with cutoff frequency ω_{C2}. with their outputs added together. The bottom diagram of Figure 18-5 shows the parallel connection of a low-pass filter and high-pass filter to form a band-reject filter.

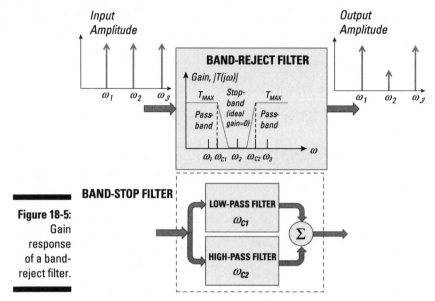

Figure 18-5:
Gain
response
of a band-
reject filter.

Make sure you select the right cutoff frequencies when you do a quick-and-dirty design of a band-reject filter based on a low-pass filter and high-pass filter connected in parallel. In Figure 18-5, if you give the low-pass filter a lower cutoff frequency of ω_{C2} and the high-pass filter an upper cutoff frequency of ω_{C1}, you'll have signals of all frequencies passing through the filter — not good for a band-reject filter. What you'll design instead is an *all-pass* filter. It's like using a coffee filter with a big, fat hole in it — everything passes through, including the coffee grounds.

Plotting Something: Showing Frequency Response à la Bode

You can express the frequency response gain $|T(j\omega)|$ in terms of decibels. Using decibels compresses the magnitude and the frequency in a logarithmic scale so you don't need more than 10 feet of paper for your plots. *Decibels* are defined as

$$|T(j\omega)|_{dB} = 20\log_{10}|T(j\omega)|$$

For example, if the gain is $|T(j\omega)| = 100$, the gain in decibels is 40 dB. Also, a gain of 1 is 0 dB.

REMEMBER

At the cutoff frequency ω_C, which is commonly defined as $T_{MAX}/\sqrt{2}$, you have the following gain:

$$\left| T(j\omega_C) \right|_{dB} = 10\log_{10}\left| \frac{T_{MAX}}{\sqrt{2}} \right|^2$$

$$= 20\log_{10}\left| \frac{T_{MAX}}{\sqrt{2}} \right| = -3 \text{ dB}$$

Therefore, the cutoff frequency is also referred to as the *–3 dB point* or the *half-power point*. Why? Because the previous set of equations involving a transfer function can be viewed as the square of either the voltage or the current transfer function. Squaring the transfer function gives you the power ratio between the output and input signal transforms because the square of the voltage or current is proportional to power. To jog your memory and give you further insight into the –3 db point as a half-power point, see Chapter 2's section on calculating the power dissipated by resistors.

The log-frequency plots of the gain $\left| T(j\omega) \right|$ and phase $\theta(\omega)$ are called *Bode plots,* or *Bode diagrams.* In the following sections, I introduce you to basic Bode plots and help you interpret them.

Looking at a basic Bode plot

Bode plots come in pairs to describe the frequency response of circuits. Usually, you have

✔ A log-frequency gain plot in decibels given in the top diagram

✔ A log-frequency phase plot in degrees given in the bottom diagram

Figure 18-6 shows a sample Bode plot.

The horizontal axis usually comes in one of the following log-frequency scales, usually decades:

✔ **Octaves:** An octave has a frequency range whose upper limit is twice the lower limit (2:1 ratio). For example, the voice usually ranges from 2 kHz to 4 kHz, spanning about 1 octave.

✔ **Decades:** A decade has a range with a 10:1 ratio. For example, human hearing usually ranges from 20 Hz to 20 kHz (20×10^3 Hz), so it spans 3 decades.

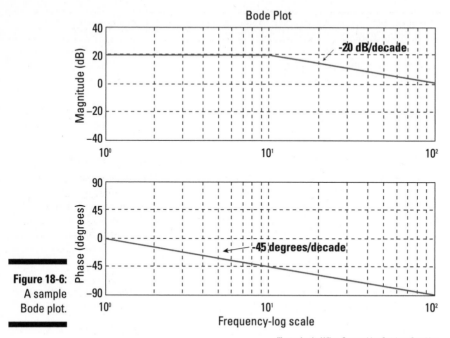

Figure 18-6:
A sample
Bode plot.

Poles, zeros, and scale factors: Picturing Bode plots from transfer functions

Most of the time, you use engineering software to draw Bode plots. But you can approximate Bode plots by hand — or at least notice when the computer-generated plot is messed up — if you understand how the transfer function's poles and zeros shape the frequency response. The *poles,* of course, are the roots of the transfer function's denominator, and *zeros* are the roots of its numerator.

Table 18-1 shows some basic, approximate rules to bear in mind when examining transfer functions and Bode plots. Figure 18-7 shows the graphical interpretation for each of the items in Table 18-1.

Table 18-1	Relating Bode Plots to a Transfer Function	
Characteristic of the Transfer Function, $T(j\omega)$	*Effects on the Gain Plot,* $\left\| T(j\omega) \right\|_{dB}$	*Effects on the Phase Plot,* $j\theta_{dB}$
Scale factor (gain)	Shifts the entire gain plot up or down without changing the cutoff (corner) frequencies	The phase Bode plot is unaffected if the scale factor is positive. If the scale factor is negative, the phase Bode plot shifts by ±180°.
Real pole	Introduces a slope of −20 dB/decade to the gain Bode plot, starting at the pole frequency	The phase Bode plot rolls off at a slope of −45°/decade. The phase at the pole is −45°. For frequencies greater than 10 times the pole frequency, the phase angle contributed by a single pole is approximately −90°.
Real zero	Introduces a slope of +20 dB/decade to the gain Bode plot, starting at the zero frequency	The phase Bode plot rolls off at a slope of +45°/decade. The phase at the zero is +45°. For frequencies greater than 10 times the zero frequency, the phase angle contributed by a single real zero is approximately +90°.
Integrator	Introduces a real pole at the origin; a real pole at the origin (an integrator 1/s) has a gain slope of −20 dB/decade passing through 0 dB at $\omega = 1$	The angle contributed by an integrator is −90° at all frequencies.

(continued)

Table 18-1 (continued)

Characteristic of the Transfer Function, $T(j\omega)$	Effects on the Gain Plot, $\lvert T(j\omega)\rvert_{dB}$	Effects on the Phase Plot, $j\theta_{dB}$
Differentiator	Introduces a real zero at the origin; a zero at the origin (a differentiator) has a gain slope of +20 dB/decade passing through at 0 dB at $\omega = 1$	The angle contributed by a differentiator is +90° at all frequencies.
Complex pair of poles	Provides a slope of –40 dB/decade	The phase Bode plot has a slope of –90°/decade. The phase at the complex pole frequency is –90°. For frequencies greater than 10 times the cutoff frequency, the phase angle contributed by a complex pair of poles is approximately –180°.
Complex pair of zeros	Provides a slope of +40 dB/decade	The phase Bode plot has a slope of +90°/decade. The phase at the complex zero frequency is +90°. For frequencies greater than 10 times the cutoff frequency, the phase angle contributed by a complex pair of zeros is approximately +180°.

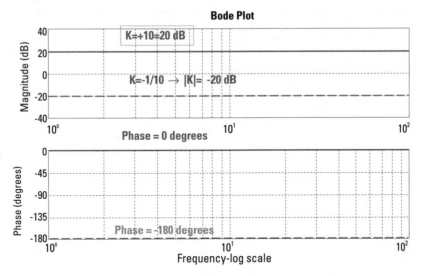

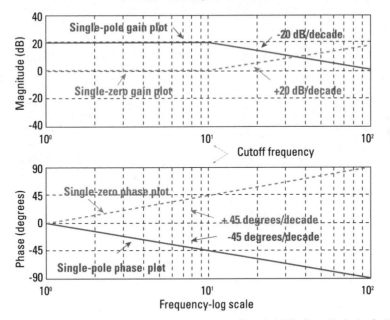

Illustration by Wiley, Composition Services Graphics

Figure 18-7: Bode diagrams of scale factors, poles, and zeros.

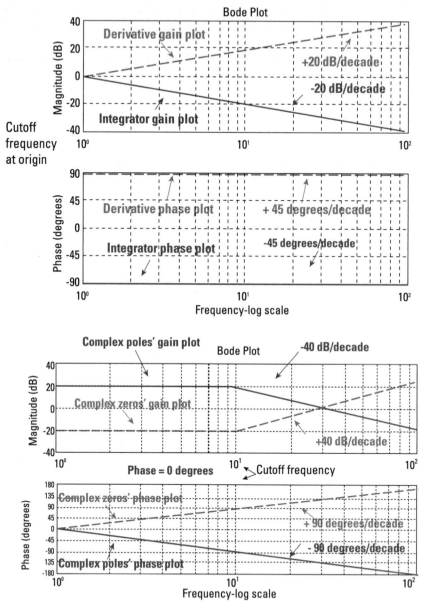

Turning the Corner: Making Low-Pass and High-Pass Filters with RC Circuits

With simple first-order circuits, you can build low-pass and high-pass filters. These simple circuits can give you a foundational understanding of how filters work so you can build more-complex filters. In the following sections, I show you how to use RC circuits to build both low-pass and high-pass filters. Later in the chapter, I show you how to build band-pass and band-reject filters based on the low-pass and high-pass filters.

First-order RC low-pass filter (LPF)

Figure 18-8 shows an RC series circuit — a circuit with a resistor and capacitor connected in series. You can get a low-pass filter by forming a transfer function as the ratio of the capacitor voltage $V_C(s)$ to the voltage source $V_S(s)$.

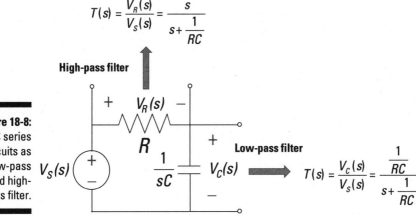

$$T(s) = \frac{V_R(s)}{V_S(s)} = \frac{s}{s + \dfrac{1}{RC}}$$

High-pass filter

Figure 18-8:
RC series circuits as a low-pass and high-pass filter.

$$T(s) = \frac{V_C(s)}{V_S(s)} = \frac{\dfrac{1}{RC}}{s + \dfrac{1}{RC}}$$

Low-pass filter

Illustration by Wiley, Composition Services Graphics

You start with the voltage divider equation:

$$V_C(s) = V_S(s) \frac{\dfrac{1}{sC}}{R + \dfrac{1}{sC}}$$

The transfer function $T(s)$ equals $V_C(s)/V_S(s)$. With some algebra (including multiplying the numerator and denominator by s/R), you get a transfer function that looks like a low-pass filter:

$$T(s) = \frac{V_C(s)}{V_S(s)} = \frac{\dfrac{1}{RC}}{s + \dfrac{1}{RC}}$$

You have a pole or corner (cutoff) frequency at s = $-1/(RC)$, and you have a DC gain of 1 at $s = 0$. The frequency response starts at $s = 0$ with a flat gain of 0 dB. When it hits $1/(RC)$, the frequency response rolls off with a slope of -20 dB/decade.

For circuits with only passive devices, you never get a gain greater than 1.

First-order RC high-pass filter (HPF)

To form a high-pass filter, you can use the same resistor and capacitor connected in series from Figure 18-8, but this time, you measure the resistor voltage $V_R(s)$. You start with the voltage divider equation for the voltage across the resistor $V_R(s)$:

$$V_R(s) = \left(\frac{R}{R + \dfrac{1}{sC}} \right) V_S(s)$$

With some algebraic manipulation (including multiplying the numerator and denominator by s/R), you can find the transfer function $T(s) = V_R(s)/V_S(s)$ of a high-pass filter:

$$T(s) = \frac{V_R(s)}{V_S(s)} = \frac{s}{s + \dfrac{1}{RC}}$$

You have a zero at $s = 0$ and a pole at $s = -1/(RC)$. You start off the frequency response with a zero with a positive slope of 20 dB/decade, and then the response flattens out starting at $1/(RC)$. You have a constant gain of 1 at high frequencies (or at infinity) starting at the pole frequency.

Creating Band-Pass and Band-Reject Filters with RLC or RC Circuits

The following sections show you how series and parallel RLC circuits form band-pass and band-reject filters. I also show you some quick-and-dirty band-pass and band-reject filters you can make using only capacitors, resistors, and op amps. These circuits come in handy when you don't have inductors lying around, though you do need an external power source to make the op amps work. These filters are built around basic RC circuits.

Getting serious with RLC series circuits

With a circuit that has a resistor, inductor, and capacitor connected in series, you can form a band-pass filter or band-reject filter.

RLC series band-pass filter (BPF)

You can get a band-pass filter with a series RLC circuit by measuring the voltage across the resistor $V_R(s)$ driven by a source $V_S(s)$. Start with the voltage divider equation:

$$V_R(s) = \left(\frac{R}{sL + \frac{1}{sC} + R} \right) V_S(s)$$

With some algebraic manipulation, you obtain the transfer function, $T(s) = V_R(s)/V_S(s)$, of a band-pass filter:

$$T(s) = \frac{V_R(s)}{V_S(s)} = \left(\frac{R}{L}\right)\frac{s}{\left[s^2 + \left(\frac{R}{L}\right)s + \frac{1}{LC}\right]}$$

Plug in $s = j\omega$ to get the frequency response $T(j\omega)$:

$$T(j\omega) = \frac{V_R(j\omega)}{V_S(j\omega)} = \left(\frac{R}{L}\right)\frac{j\omega}{\left[(j\omega)^2 + \left(\frac{R}{L}\right)j\omega + \frac{1}{LC}\right]}$$

$$= \left(\frac{R}{L}\right)\frac{j\omega}{\left[\left(\frac{1}{LC} - \omega^2\right) + \left(\frac{R}{L}\right)j\omega\right]}$$

The $T(j\omega)$ reaches a maximum when the denominator is a minimum, which occurs when the real part in the denominator equals 0. In math terms, this means that

$$\frac{1}{LC} - \omega^2 = 0 \quad \rightarrow \quad \omega_0 = \frac{1}{\sqrt{LC}}$$

The frequency ω_0 is called the *center frequency.*

The cutoff frequencies are at the –3 dB half-power points. The –3 dB point occurs when the real part in the denominator is equal to $R\omega/L$:

$$\frac{1}{LC} - \omega^2 = \pm\frac{R}{L}\omega \quad \rightarrow \quad \omega^2 \pm \frac{R}{L}\omega - \frac{1}{LC} = 0$$

You basically have a quadratic equation, which has four roots due to the plus-or-minus sign in the second term. The two appropriate roots of this equation give you cutoff frequencies at ω_{C1} an ω_{C2}:

$$\omega_{C1} = -\frac{R}{2L} + \sqrt{\left(\frac{R}{2L}\right)^2 + \frac{1}{LC}}$$

$$\omega_{C2} = \frac{R}{2L} + \sqrt{\left(\frac{R}{2L}\right)^2 + \frac{1}{LC}}$$

The *bandwidth BW* defines the range of frequencies that pass through the filter relatively unaffected. Mathematically, it's defined as

$$BW = \omega_{C2} - \omega_{C1} = \frac{R}{L}$$

Another measure of how narrow or wide the filter is with respect to the center frequency is the *quality factor Q*. The quality factor is defined as the ratio of the center frequency to the bandwidth:

$$Q = \frac{\omega_0}{BW} = \frac{1/\sqrt{LC}}{R/L}$$

$$= \frac{1}{R}\sqrt{\frac{L}{C}}$$

The RLC series circuit is *narrowband* when $Q \gg 1$ (high Q) and *wideband* when $Q \ll 1$ (low Q). The separation between the narrowband and wideband responses occurs at $Q = 1$. Figure 18-9 shows the series band-pass circuit and gain equation for an RLC series circuit.

$$T(s) = \frac{V_0(s)}{V_s(s)} = \frac{s^2 + \frac{1}{LC}}{s^2 + \left(\frac{R}{L}\right)s + \frac{1}{LC}}$$

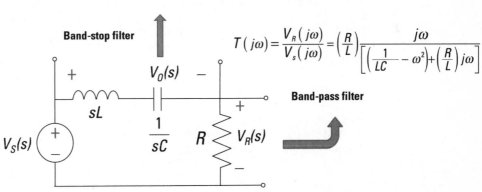

Figure 18-9: An RLC series circuit as a band-pass filter and a band-reject filter.

$$T(j\omega) = \frac{V_R(j\omega)}{V_s(j\omega)} = \left(\frac{R}{L}\right)\frac{j\omega}{\left[\left(\frac{1}{LC} - \omega^2\right) + \left(\frac{R}{L}\right)j\omega\right]}$$

Illustration by Wiley, Composition Services Graphics

The frequency response is shaped by poles and zeros. For this band-pass filter, you have a zero at $\omega = 0$. You start with a gain slope of +20 dB. You hit a cutoff frequency at ω_{C1}, which flattens the frequency response until you hit another cutoff frequency above ω_{C2}, resulting in a slope of –20 dB/decade.

RLC series band-reject filter (BRF)

You form a band-reject filter by measuring the output across the series connection of the capacitor and inductor. You start with the voltage divider equation for the voltage across the series connection of the inductor and capacitor:

$$V_0(s) = \left(\frac{sL + \dfrac{1}{sC}}{sL + \dfrac{1}{sC} + R} \right) V_S(s)$$

You can rearrange the equation with some algebra to form the transfer function of a band-reject filter:

$$T(s) = \frac{V_0(s)}{V_S(s)} = \frac{s^2 + \dfrac{1}{LC}}{s^2 + \left(\dfrac{R}{L}\right)s + \dfrac{1}{LC}}$$

When you plug in $s = j\omega$, you have poles and zeros shaping the frequency response. For the band-reject filter, you have a double zero at $1/\sqrt{LC}$. Starting at $\omega = 0$, you have a gain of 0 dB. You hit a pole at ω_{C1}, which rolls off at –20 dB/decade until you hit a double zero, resulting in a net slope of +20 dB/decade. The frequency response then flattens out to a gain of 0 dB at the cutoff frequency ω_{C2}. You see how the poles and zeros form a band-reject filter.

Climbing the ladder with RLC parallel circuits

You can get a transfer function for a band-pass filter with a parallel RLC circuit, like the one in Figure 18-10.

$$T(j\omega) = \frac{I_R(s)}{I_s(s)} = \left(\frac{1}{RC}\right)\frac{j\omega}{\left[\left(\frac{1}{LC} - \omega^2\right) + \left(\frac{1}{RC}\right)j\omega\right]}$$

Band-pass filter

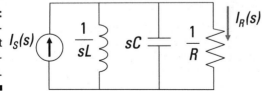

Figure 18-10:
An RLC parallel circuit as a band-pass filter.

Illustration by Wiley, Composition Services Graphics

You can use current division to find the current transfer function of the parallel RLC circuit. By measuring the current through the resistor $I_R(s)$, you form a band-pass filter. Start with the current divider equation:

$$I_R(s) = \left(\frac{\frac{1}{R}}{sC + \frac{1}{sL} + \frac{1}{R}}\right)I_s(s)$$

A little algebraic manipulation gives you a current transfer function, $T(s) = I_R(s)/I_s(s)$, for the band-pass filter:

$$T(s) = \frac{I_R(s)}{I_s(s)} = \left(\frac{1}{RC}\right)\left(\frac{s}{s^2 + \left(\frac{1}{RC}\right)s + \frac{1}{LC}}\right)$$

Plug in $s = j\omega$ to get the frequency response $T(j\omega)$:

$$T(j\omega) = \frac{I_R(j\omega)}{I_s(j\omega)} = \left(\frac{1}{RC}\right)\frac{j\omega}{\left[\left(\frac{1}{LC} - \omega^2\right) + \left(\frac{1}{RC}\right)j\omega\right]}$$

This equation has the same form as the RLC series equations (see the earlier section "Getting serious with RLC series circuits" for details). For the rest of this problem, you follow the same process as for the RLC series circuit.

The transfer function is at a maximum when the denominator is minimized, which occurs when the real part of the denominator is set to 0. The cutoff frequencies are found when their gains $|T(j\omega_c)| = 0.707|T(j\omega_0)|$ or the –3 dB point. Therefore, ω_0 is

$$\frac{1}{LC} - \omega^2 = 0 \quad \rightarrow \quad \omega_0 = \frac{1}{\sqrt{LC}}$$

The center frequency, the cutoff frequencies, and the bandwidth have equations indentical to the ones for the RLC series band-pass filter.

Your cutoff frequencies are ω_{C1} and ω_{C2}:

$$\omega_{C1} = -\frac{1}{2RC} + \sqrt{\left(\frac{1}{2RC}\right)^2 + \frac{1}{LC}}$$

$$\omega_{C2} = \frac{1}{2RC} + \sqrt{\left(\frac{1}{2RC}\right)^2 + \frac{1}{LC}}$$

The bandwidth BW and quality factor Q are

$$BW = \omega_{C2} - \omega_{C1} = \frac{1}{RC}$$

$$Q = \frac{\omega_0}{BW} = R\sqrt{\frac{C}{L}}$$

RC only: Getting a pass with a band-pass and band-reject filter

Using simple first-order low-pass and high-pass filters based on the RC series circuit, you can form quick-and-dirty band-pass and band-reject filters with gain. You use a noninverting amplifier filter to provide circuit isolation between the low-pass filter and the high-pass filter.

The top diagram in Figure 18-11 shows this technique for a band-pass filter. The dashed lines indicate the RC series low-pass filter and high-pass filter and the noninverting amplifier. To form a band-reject filter, you can take the outputs of an RC series low-pass filter and high-pass filter with an inverting adder. The bottom of Figure 18-11 points out the key components of the quick-and-dirty band-reject filter: a low-pass filter, a high-pass filter, and an inverting adder. For this band-reject filter design, you need to choose values to prevent loading effects.

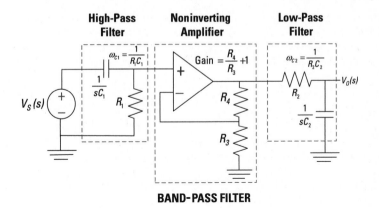

BAND-PASS FILTER

Figure 18-11: Quick-and-dirty band-pass and band-reject filters.

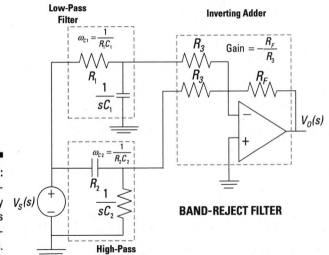

BAND-REJECT FILTER

Illustration by Wiley, Composition Services Graphics

Part VI
The Part of Tens

Visit www.dummies.com/extras/circuitanalysis for the scoop on ten mistakes commonly made in circuit analysis.

In this part . . .

✔ Survey ten practical applications for circuits.

✔ Take a look at ten technologies affecting circuits.

Chapter 19

Ten Practical Applications for Circuits

. . . .

In This Chapter

▶ Examining variable resistors and homemade capacitors

▶ Considering interface techniques and variations on the Wheatstone bridge

. . . .

*P*art of the purpose of circuit analysis is to analyze what a circuit is doing. But you can also use circuit analysis to design a circuit to perform a particular function. Knowing how to analyze circuits allows you to add the appropriate elements to a circuit during the design phase so that the circuit performs the way you want it to. In this chapter, I highlight ten of my favorite practical applications for circuits.

Potentiometers

Dimmer switches are actually adjustable voltage dividers referred to as *potentiometers* in the electrical engineering world. From a circuit analysis perspective, you can model a potentiometer as two resistors connected in series. The connection or junction point between the two resistors is where the wiper arm is located to vary the resistance. By varying the amount of resistance, you vary the amount of voltage. In the case of dimmer switches, this variance allows you to adjust the lighting in a room.

Homemade Capacitors: Leyden Jars

Hailing from Holland more than 250 years ago is a simple capacitor known as a Leyden jar. It was a major breakthrough, replacing insulated conductors of large dimensions to store charge. A Leyden jar consists of one piece of metal foil coating the inside of a glass jar and another piece of metal foil coating the outside. The jar serves as the dielectric insulator between the two conducting foils. The Leyden jar holds electricity where energy is stored within the glass.

Digital-to-Analog Conversion Using Op Amps

To talk to the real world, a computer needs digital-to-analog converters. You can use an operational amplifier (op amp) with multiple inputs to feed an inverting summer using an op amp. To reduce the number of resistors using an inverting summer, you use an R-2R network. Only two resistor values are needed in an R-2R network for any number of digital inputs. When analyzing this circuit, you use the superposition concepts in Chapter 7 and the Thévenin equivalent, which I explain in Chapter 8.

Two-Speaker Systems

A two-speaker system has one speaker called a *tweeter* that handles high-frequency music and another speaker called the *woofer* that handles the low frequencies. The input audio signal feeds across the series connection of a capacitor and resistor. The resistor terminals feed the input of the tweeter, and the capacitor terminals feed the woofer. To analyze a two-speaker system from a frequency perspective, you need to know about high-pass and low-pass filters (see Chapter 18).

Interface Techniques Using Resistors

You can connect resistors to a device load in order to avoid exceeding the device's power ratings. For example, you can connect a resistor in series or in parallel to a device wherever you want to limit the voltage across and/or the current through the device.

Say you want to show that a light emitting diode (LED) is turned on. You need to limit the current going through the LED; otherwise, you may destroy it with too much current. To limit the current, you connect resistors to limit the current or voltage of the diode. Circuit analysis helps you determine how much resistance you need to protect the diode.

Interface Techniques Using Op Amps

You can take a physical variable such as temperature range and convert it to a voltage range. For example, suppose you need an amplifier to pump up a weak signal from a temperature transducer. (A *transducer* converts a physical variable to an electrical variable.) The amplifier's output is fed to a two-input summer with gain. The other input is a constant voltage source that moves the signal up and down the desired voltage range via a potentiometer (variable resistor).

The Wheatstone Bridge

The Wheatstone bridge is a circuit used to measure unknown resistances. Mechanical and civil engineers measure resistances of strain gauges to find the stress and strain in machines and buildings. The bridge network has three precision resistors and one unknown resistor. Two of the known resistors are potentiometers, which are adjusted to balance the bridge network and thus determine the unknown resistor. The accelerometer, which I describe in the next section, uses a Wheatstone bridge arrangement of strain gauges.

Accelerometers

You can use an array of strain gauges to develop accelerometers. A common strain gauge consists of a flexible backing that supports a metallic foil pattern. The strain gauge leverages the changes of its physical dimensions when mechanical force acts on the strain gauge.

Suppose that inside a rocket, you have a miniaturized cantilever beam with a mass hanging at one end. You've placed a pair of strain gauges at the top and another pair at the bottom of the cantilever beam. When there's an upward accelerated force, the cantilever beam bends downward. The strain gauges on top stretch from the upward acceleration, resulting in an increase in resistance; meanwhile, the strain gauges at the bottom get compressed, decreasing the resistance of the strain gage. Because of the difference of resistances in the strain gauges, you can use voltage divider arrangements and a bridge network (called a Wheatstone bridge; see the preceding section) to determine the amount of acceleration.

Electronic Stud Finders

You can modify the Wheatstone bridge by replacing two of the four resistors with two capacitors and setting the other known resistors equal. You have two resistor-capacitor (RC) series branches connected in parallel fed by an AC source. You take each point between the RC series combination and feed it to an audio differential amplifier connected to a beeper. The capacitances between the metal plates vary as the plates pass over the stud. The capacitors are equal when the stud finder is centered on the stud, and the Wheatstone bridge is said to be balanced. When the stud is off-center, the capacitors are unequal and a sound is emitted.

555 Timer Circuits

Invented in 1971, the 555 timer chip remains a popular integrated circuit among hobbyists. You can use an external resistor and capacitor network to change the timing interval by the careful choice of resistors and capacitors.

You can configure the 555 chip to work like a cooking timer (a one shot). After you set the timer and a certain amount of time elapses, the timer goes off. Or you can configure the 555 chip as a two-state clock (astable), triggering a series of pulses at regular intervals. Other applications of a 555-timer include a Morse code call-sign generator, a metronome circuit, and an alarm circuit when your vehicle's windshield wiper fluid runs low.

Analyzing the circuitry of the 555 chip may require you to review Chapter 13 on first-order circuits to see how timing works when using capacitors.

Chapter 20

Ten Technologies
Affecting Circuits

In This Chapter
▶ Discovering technologies affecting circuit complexity and analysis
▶ Addressing the future impact on various industries

Circuit analysis involves designing new circuits as emerging technologies mature and become commonplace. And of course, integrating all the components of these new technologies requires circuit analysis. This chapter lists ten exciting technologies used in current and up-and-coming circuits.

Smartphone Touchscreens

The touchscreens found on smartphones use a layer of capacitive material to hold an electrical charge; touching the screen changes the amount of charge at a specific point of contact. In resistive screens, the pressure from your finger causes conductive and resistive layers of circuitry to touch each other, changing the circuits' resistance. When you locate the capacitance or resistance changes with a coordinate system, you can have multiple fingers controlling the display of the smartphone.

Nanotechnology

In nanotechnology, devices operate at the molecular scale, between 1 and 100 nanometers (the Greek prefix *nano-* means one-billionth, or 10^{-9}). Research in nanotechnology developed techniques to design and build electronic devices and mechanical structures with atomic-level control. With atomic-level control, you can synthesize materials with optimum properties, such as resistance and material strength. With the circuit size reduced, the system speed increases, and it's possible to operate devices within the terahertz (10^{12} Hz) range.

Nanotechnology offers a lot of promise in a variety of fields. It may reduce greenhouse gases, limit deforestation, decrease pollution, and allow cheap manufacturing. For the home, high-tech gadgets may identify deadly bacteria. On the medical front, small, inexpensive implantable sensors could monitor your health and provide semiautomated treatment.

Carbon Nanotubes

One special category of nanotechnology is the use of carbon nanotubes. Carbon nanotubes are hollow structures with walls formed by one-atom-thick sheets of carbon. The sheets are rolled at specific discrete angles to determine the nanotube properties, such as strength. Carbon nanotubes have a wide range of potential applications. Here are just a few:

- **Targeted medication:** Coating porous plastic with carbon nanotubes can create implantable biocapsules that can detect problems in blood chemistry and, for example, release insulin for persons with diabetes. Or such capsules could deliver chemotherapy drugs directly to thermal cells.

- **Cleaning oil spills:** When you add boron atoms to growing carbon nanotubes, the nanotubes become sponge-like, absorbing oil.

- **Creating new materials:** Carbon nanotubes may be used to create new materials that change the surface shapes of aircraft wings when a voltage is applied. Carbon nanotubes can also fill voids found in conventional concrete, preventing water from entering the concrete and increasing the concrete's lifetime.

- **Energy efficiency:** You can recycle wasted heat as electricity by using thermocells that use nanotube electrodes.

Microelectromechanical Systems

Microelectromechanical systems (MEMS) devices are manufactured using similar microfabrication techniques as those used to build integrated circuits. MEMS can have moving components that allow the device to perform physical or analytical functions in addition to the electrical function. The vast assortment of MEMS components and devices built and tested in the laboratory environment includes micropumps, cantilevers, rotors, channels, valves, and sensors. In biomedical applications, MEMS can be used for retinal implants to treat blindness, neural implants for stimulation and recording from the central nervous system, and microneedles for painless vaccinations.

The digital aspects of MEMS, along with built-in optical, electrical, or chemical sensing components, may provide timely delivery of drugs when human cells start to get sick. Due to the short time scale under physiologically relevant conditions, MEMS can activate body systems by delivering an electrical stimulus, drugs, or both.

Supercapacitors

Supercapacitors (or supercaps) are energy storage devices with very high capacity and low internal resistance. The energy is stored in a double-layer electrolytic material, so supercaps are often called *electrochemical double-layer capacitors* (EDLC) — I like the term *supercaps* better. Compared to conventional electrolytic capacitors, supercaps have high energy and power densities, as well as a longer lifetime. And as of 2011, the capacitance could go up to several thousand farads.

The Memristor

You know the capacitor, resistor, and inductor are the three basic circuit elements. In 1971, Leon Chua postulated a fourth basic circuit element named a memristor (memory and resistor). Unlike the other three elements, the hypothetical memristor can memorize the nonlinear resistance by controlling the charge or magnetic flux. Unlike conventional resistors, the direct current (DC) resistance of the memristor depends on the total charge that passes through the device in a given time interval. If you turn off the driving signal, the memristor's resistance stays at that value until the signal is turned back on.

Because of the nonvolatile properties, the memristor could be used in high-density storage devices. Other possible applications for memristors include reprogrammable digital logic circuits and smart interconnects.

Superconducting Digital Electronics

Digital semiconductor devices have been shrinking in size for many decades. As you shrink these devices, heating becomes an important problem along with increasing delay times due to wire (trace) resistance. Superconducting digital devices offer high speed and reduced power with high-density packaging and superconducting interconnects. Power consumption in high-frequency operation is three orders of magnitude less than the CMOS (complementary metal oxide semiconductor) logic, which is a type of circuitry that minimizes the amount of power used.

Wide Bandgap Semiconductors

Wide bandgap materials are semiconductors with bandgaps greater than 1 electron volt (eV). Wide bandgap semiconductors such as silicon carbide (SiC) and gallium nitride (GaN) promise to revolutionize both optoelectronic and electronic devices. New lasers and light emitting diodes (LEDs) are possible, including blue-green lasers, blue-green or white LEDs, solar-blind detectors, high-power solid-state switches and rectifiers, and high-power microwave transistors. Another possible role of wide bandgap semiconductors is high-temperature electronics, especially in automotive, aerospace, and energy applications. High-temperature semiconductors must operate higher than 150 degrees Celsius (about 300 degrees Fahrenheit).

Flexible Electronics

Flexible electronics covers a wide range of device and materials technologies that are built on flexible and *conformal substrates* (substrates that conform to the shape of a flexible surface so you can imprint electronic components). They provide opportunities to integrate a variety of components that are fluidic, mechanical, optical, and electronic.

Radio frequency identification (RFID) tagging has emerged as one of the building blocks of flexible electronics. Other technologies, including carbon nanotubes (see the earlier section "Carbon Nanotubes"), nanowires, and other nanomaterials within semiconductors are being developed to tailor properties of cost, mobility, and scalability. Flexible electronics may have additional applications in healthcare, the automotive industry, human-machine interactivity, energy management and mobile devices, wireless systems, and electronics embedded in living and hostile environments.

Microelectronic Chips that Pair Up with Biological Cells

By growing biological cells atop CMOS-based microelectrode arrays, researchers can study — and emulate — how information is processed in the brain. How does this work? The answer lies in the rich connectivity and highly coordinated electrical signals coming from neural cells and neural networks.

By adopting integrated circuit (IC) or CMOS (complimentary metal oxide semiconductor) technology, you can address the connectivity of many transducers or electrodes by using automated electronics to look at an array of sensors or transducers; condition the signal quality at the electrode using dedicated circuitry such as filters and amplifiers; and reduce the system complexity, because many functions can be programmed through software and digital registers on the chip side.

Major challenges posed by using CMOS technology are protecting and packaging the chip so that it doesn't corrode and poison the cell.

Index

• A •

AC (alternating current), 8–9
accelerometers, 339–340
active circuits, 134
active devices, 12, 134
adder (summing amplifier), 164–166, 172, 210
additive property, 98, 258. *See also* superposition
admittance, 260, 265, 299
alternating current (AC), 8–9
amperes, 8
amplifying. *See also* operational amplifiers (op amps)
  current, 146–149
  signals, 149–151

• B •

band-pass filters
  bandwidth of, 329
  cutoff frequencies of, 317, 328, 332
  frequency response of, 316, 317, 330
  narrowband versus wideband, 329
  quality factor of, 329
  RC series, 332–333
  RLC parallel, 330–332
  RLC series, 327–330
  transfer function for, 328, 330–332
band-reject filters
  cutoff frequencies of, 317, 318
  frequency response of, 317, 318, 330
  RC series, 332–333
  RLC series, 330
  transfer function for, 330
bandwidth, 329
base terminal of bipolar transistors, 145

batteries, 18–19
bipolar transistors
  amplifying current with, 146–149
  amplifying signals with, 149–151
  common base circuit, 145, 146, 149–151
  common collector circuit, 145, 146, 151–153
  common emitter circuit, 145, 146–149
  current buffer, 149
  emitter follower, 151
  isolating circuits with, 151–153
  NPN versus PNP, 146
  origin of name, 145
  terminals of, 145
birds on high-voltage lines, 19
Bode plots or diagrams
  basic, 319–320
  log-frequency scales in, 319
  relating to transfer function characteristics, 320–324
bombs, as impulse functions, 176

• C •

capacitance, 194, 199–200, 212
capacitors
  charging and discharging, 195
  circuit symbols of, 194
  current-voltage relationship, 195–196
  defining equation for, 194
  described, 194
  element constraint for, 216, 238, 286, 292–293, 299
  finding the power of, 196–198
  in first-order circuits, 13, 212
  homemade (Leyden jars), 338
  impedance and admittance in the *s*-domain, 299

capacitors *(continued)*
  impedance replacing, 264
  i-v relationships in the *s*-domain,
    297–298
  Ohm's law in phasor form for, 262
  in op-amp integrator, 205–207
  output values for RC time constants, 224
  parallel model in the *s*-domain, 298
  phasor diagram of, 261
  in RC series circuits, 212
  in second-order circuits, 13
  series model in the *s*-domain, 298
  supercapacitors, 343
  uses for, 193–194
carbon nanotubes, 342
CCCS (current-controlled current
    source), 134–135, 146–147, 297
CCVS (current-controlled voltage
    source), 134–135, 297
characteristic equation
  for first-order differential equations,
    214–215
  for RC series circuit, 218–219
  for RL parallel circuit, 227–228
  for RLC parallel circuit, 249
  for RLC series circuit, 240
  for second-order differential
    equations, 236
charge, 8
charging capacitors, 195
Chua, Leon (theorist), 343
CMOS (complimentary metal oxide
    semiconductor) technology, 344
collector terminal of bipolar
    transistors, 145
common base circuit, 145, 146, 149–151
common collector circuit, 145, 146,
    151–153
common emitter circuit, 145, 146–149
complex frequency variable, 274
complex numbers. *See also* phasors;
    imaginary numbers; impedance;
    s-domain
  exponential, 257, 259
  Euler's formula, 257

  defined, 252
  poles, 282, 284, 278–280, 322
  roots, 240, 242, 246, 249, 282
  zeros, 282, 322
complimentary metal oxide
    semiconductor (CMOS)
    technology, 344
conductance, 17, 37–38, 40
connection equations, 25
conservation laws, 25
conservation of energy in KVL equations,
    28–29
constants
  capitalization conventions for, 1
  proportionality, 135, 260
  time, 219, 228, 231
converting circuit sources, 45–49
cosine function, 257, 275
coulombs per second, 8, 16
current. *See also* Kirchhoff's current law
    (KCL); mesh currents
  alternating (AC) versus direct (DC), 8–9
  amplifying with common emitter
    circuit, 146–151
  of capacitors, relationship of voltage to,
    195–196
  connection equations for, 32–33
  constant with infinite resistance, 20
  coulombs per second as units of, 8, 16
  defined, 8
  device equations for, 31, 33–34, 76, 77
  device, mesh currents' relationship to,
    84–86
  direction of, 10
  in electrical power function, 15
  limited by resistors, 16
  in Ohm's law, 17
  one point needed to measure, 8
  positive versus negative, 10
  in series circuits, 31, 34–35
  as a through variable, 8
  variable for, 8
  water flow analogy for, 29
current buffer, 149

current division
  current transfer function, 58–59
  finding currents with multiple voltage sources, 60–63
  finding the current through each device in parallel circuits, 59–60
  origins of, 57–59
  for parallel circuit in the _s_-domain, 302–303
  as proportionality concept, 97, 98
  for RLC parallel band-pass filter, 331
  using repeatedly, 63–64
current, incoming and outgoing. _See_ Kirchhoff's current law (KCL)
current sources, independent. _See also_ dependent sources
  converting to and from voltage sources, 45–47
  ideal, 20
  multiple, with parallel resistors, voltage division for, 52–55
  node-voltage analysis for, 74–79
  Norton (short circuit), 116, 118, 121
  relationship between dependent and independent sources, 136
  replacing with open circuits, 102–103
  superposition with one voltage source and one current source, 100–101, 107–108
  superposition with three or more sources, 108–111
  superposition with two sources, 105–106
  turning on and off, 102–103, 136
current transfer function, 58–59
current-controlled current source (CCCS), 134–135, 146–147, 297
current-controlled voltage source (CCVS), 134–135, 297
cutoff frequencies of filters
  band-pass, 317, 328, 332
  band-reject, 317, 318
  defined, 315
  expressed in decibels, 319
  gain at, 319
  high-pass, 316
  low-pass, 315, 326

• D •

DAC (digital-to-analog conversion), 172, 338
damped cosine function, 275, 284, 285
damped functions, Laplace transform pairs for, 275
DC (direct current), 8–9
decade log-frequency scales, 319
decibels, gain in, 318–319
dependent sources. _See also specific types_
  bipolar transistors with, 145–153
  defined, 134
  JFET transistors with, 142–144
  node-voltage analysis of, 137–138
  op-amp model with, 158–159
  proportionality of output of, 135
  relationship to independent sources, 136
  source transformation with, 138–140
  Thévenin's theorem with, 140–142
  types of, 134–135
differential amplifier or subtractor, 166–168
differential equations. _See also_ first-order differential equations; second-order differential equations
  first-order versus second-order, 13
  further information, 211
  solving with op amps, 208–210
  techniques for avoiding, 13
Differential Equations For Dummies (Holzner), 211
differentiation property, 276
differentiator op-amp circuit, 207–208
digital-to-analog conversion (DAC), 172, 338
dimmer switches, 337
Dirac delta function, 176–179, 275
direct current (DC), 8–9
discharging capacitors, 195

division techniques. *See also* current division; voltage division
  in phasor form, 264–266
  proportionality use by, 97–98
drain of JFET transistor, 143
duality, 246, 249, 250, 251

• *E* •

electrical charge. *See* charge
electrical power. *See* power
electrochemical double-layer capacitors (EDLC), 343
electron, charge of, 8
electronic stud finders, 340
Electronics All-in-One For Dummies (Lowe), 145
emitter follower, 151
emitter terminal of bipolar transistors, 145
equivalent circuits, finding, 42–44. *See also* Norton's theorem; Thévenin's theorem
Euler's formula, 190–191, 257
exponential function
  connecting sinusoidal functions via Euler's formula, 190–191
  decaying exponential version, 185
  defined, 184–185
  growing exponential version, 185
  Laplace transform pairs for, 275
  poles and zeros of, 283–284
  pole-zero diagram for, 285
  for solving first-order differential equations, 213–214
  for solving second-order differential equations, 235
  time-shifted version of decaying exponential, 185

• *F* •

farads, 194
FETs (field-effect transistors), 142–144

filters
  band-pass, 316–317, 327–333
  band-reject, 317–318, 330, 332–333
  cutoff frequency of, 315, 316, 317, 318, 319
  high-pass, 316, 325, 326
  low-pass, 315, 325–326
  passband region, 314
  resistors for creating, 16
  sinusoidal steady-state output of, 314
  stopband region, 314
  transition region for nonideal, 315
  types of, 14
first-order circuits. *See also specific types*
  described, 13
  finding the total response, 213, 217, 222–224, 230–231
  finding the zero-input response, 212, 217–219, 226–228
  finding the zero-state response, 213–214, 219–222, 228–230
  initial condition needed for analysis, 212
  RC series, 212, 215–224, 286–288, 325–326, 332–333
  RL parallel, 212, 224–231, 290–292
  single storage element in, 211
first-order differential equations
  characteristic equation for, 214–215, 218–219, 227–228
  with constant coefficients, steps for solving, 212–213
  converting to algebraic equations, 214–215
  exponential function for solving, 213–214
  homogeneous solution for, 213
  i-v relationship for inductor, 212
  needed for first-order circuits, 211, 212
  Ohm's law for capacitance, 212
  particular solution for, 213
  for RC series circuit, 216, 217–218, 220, 221, 222
  for RL parallel circuit, 225, 226, 227, 228–229

555 timer circuits, 340
flexible electronics, 344
Fourier coefficients of sinusoidal
    functions, 190
frequency response. *See also* filters
  of band-pass filters, 316, 317, 330
  of band-reject filters, 317, 318, 330
  Bode plots or diagrams of, 319–324
  of high-pass filters, 316
  of low-pass filters, 315
  plotting using gain and phase, 314
  uses for, 313
F(*s*) function. *See* Laplace transform
functions. *See also* exponential function;
    transfer function
  cosine, 257, 275
  current transfer, 58–59
  damped cosine, 275, 284, 285
  impulse or Dirac delta, 176–179
  Laplace transform pairs for (table), 275
  pole-zero examples for, 283–285
  ramp, 182–184, 275, 284
  sawtooth, 184
  sine, 275, 284, 285
  sinusoidal, 186–191
  step, 177, 180–184, 275
  triangle, 183–184

• _G_ •

gain
  for active and passive circuits, 134
  amplification with active devices, 134
  Bode plots or diagrams of, 319–324
  at cutoff frequency of filters, 319
  in decibels, 318–319
  in frequency response plot, 314
  of linear dependent sources, 135
  other terms for, 134
  proportionality constant, 135, 260
  transfer function relationship, 314
gain response, 315, 316, 317, 318. *See
    also* frequency response
gates, 21, 22, 143

ground, 9
ground point. *See* reference point

• _H_ •

henries, 201
high-pass filters, 316, 325, 326
Holzner, Steven (*Differential Equations
    For Dummies*), 211
homogeneous solution
  for first-order differential equations, 213
  for RC series circuit, 220
  for RL parallel circuit, 228–229
  for RLC parallel circuit, 251
  for RLC series circuit, 243
  for second-order differential
    equations, 235

• _1_ •

icons in this book, explained, 4
ICs (integrated circuits), 133, 155, 340
imaginary numbers. *See also* complex
    numbers; phasors; *s*-domain
  Euler's formula, 257
  frequency variable s, 274
  reactance (part of impedance), 260
  roots or poles, 282, 284
  rotating phase angle, 259, 262, 263
impedance
  admittance (*Y*) as reciprocal of, 260
  as complex number, 260
  equivalent for RLC parallel circuit, 265
  equivalent for RLC series circuit, 264
  Ohm's law in terms of, 260
  as a proportionality constant, 260
  real and imaginary parts of, 260
  replacing resistors, inductors, and
    capacitors with, 264
  in the *s*-domain, 299
  with superposition in phasor domain,
    266–267
  Thévenin and Norton equivalents,
    268–270

impulse function, 176–179, 275
independent equations, 32
independent sources. *See* current
  sources, independent; voltage
  sources, independent
inductance
  finding for parallel inductors, 204–205
  finding for series inductors, 203–204
  in inductor equation, 201
inductors
  circuit symbol of, 201
  current through, 202
  defining equation for, 201
  described, 201
  element constraint for, 225, 238, 290,
    292, 299
  finding the energy storage of, 202–203
  in first-order circuits, 13, 212
  impedance and admittance in the
    s-domain, 299
  impedance replacing, 264
  i-v characteristics of, 201, 202, 212
  i-v relationships in the *s*-domain,
    297–298
  linear relationship of, 202
  Ohm's law in phasor form for, 263
  parallel model in the *s*-domain, 298
  phasor diagram of, 261
  in RL parallel circuits, 212
  in second-order circuits, 13
  series model in the *s*-domain, 298
  uses for, 193, 200
infinite resistance, 20, 102
initial condition, 212. *See also* zero-input
  response calculation; zero-state
  response calculation
instrumentation amplifier, 168–169
integrated circuits (ICs), 133, 155, 340
integration property, 275
integrator op-amp circuit, 205–207, 210
integro-differential equation, 293–294
interface techniques, 338–339
inverse Laplace transform, applying, 274.
    See also partial fraction expansion

inverting op amps
  analysis of, 163–164
  in differential amplifier or subtractor,
    166–168
  integrator, 205–207, 210
  summer or summing amplifier,
    164–166, 210
  voltage division with, 161
isolating circuits, 151–153
i-v characteristics
  of batteries, 18
  of circuits with only resistors and
    independent power sources, 16
  finding, 43–44
  of inductors, 201, 202, 212
  for RC series circuit devices, 216, 286
  of resistors, 16, 17
  for RL parallel circuit devices,
    225, 290
  same in equivalent circuits, 42–44
  of short circuits, 18
  in superposition, 98–99, 101–103

• *J* •

j. See imaginary numbers
JFETs (junction field-effect transistors),
  142–144
joules per coulomb, 16
joules per second, 16

• *K* •

Kirchhoff, Gustav, 25
Kirchhoff's current law (KCL)
  calculating, 30
  connection equations using, 31, 32–33
  defined, 29
  described, 11, 29
  finding *i-v* characteristics using, 43–44
  finding short-circuit current with,
    120–121
  forming an equation, 29–30
  inverting op amp analysis using, 163–164

in node-voltage analysis, 69, 70, 75, 78, 81, 137–138, 303–304
number of equations per circuit, 30
procedure for circuit analysis, 31
in the *s*-domain, 296
Kirchhoff's voltage law (KVL)
circuit diagram illustrating, 27–28
connection equations using, 31, 32
as conservation of energy equation, 28–29
defined, 26
described, 11, 26
finding *i-v* characteristics using, 43
finding open-circuit voltage with, 119–120
forming an equation, 27–29
in mesh-current analysis, 86, 87
parallel circuit analysis with, 36–37
procedure for circuit analysis, 31
in the *s*-domain, 296
voltage division using, 50

• *L* •

lag signal, 187–188
Laplace domain. *See s*-domain
Laplace transform
described, 14
differentiation property with, 276
F(*s*) function in *s*-domain found by, 274
integration property with, 275
key pairs (table), 275
linearity property with, 275
partial fraction expansion, simple version, 276–278
partial fraction expansion, with complex poles, 278–280
partial fraction expansion, with multiple poles, 280–282
pole-zero diagram for, 282–283
pole-zero examples for functions, 283–285
process of applying, 274

as ratios of polynomials in the *s*-domain, 282
for RC series circuit, 286–288
for RL parallel circuit, 290–292
for RLC series circuit, 292–294
steps for analyzing a circuit using, 285
lead signal, 187–188
Leyden jars, 338
lightning, as an impulse function, 176
linear resistors' *i-v* characteristics, 17
linearity
additive property of, 98
in Laplace transform approach, 275
proportionality property of, 96–98
load circuit, 113, 114, 127–129. *See also* Norton's theorem; Thévenin's theorem
loading, avoiding, 162–163
log-frequency scales, 319
loops, 22–23, 27–28
low-pass filters, 315, 325–326
L/R time constant, 228, 231

• *M* •

magnetic flux, 201
makeover techniques. *See* source transformation
matrix form for equations, 72–73, 89
maximum power, delivering to a load circuit, 127–129
maximum power theorem, 127
memristor, 343
mesh currents, 84–86
mesh-current analysis
for common emitter circuit, 147–148
described, 12
KVL equations for, 86, 87
matrix form for equations, 89
Ohm's law in, 87–88
in phasor domain, 271–272
in the *s*-domain, 304–305
solving for unknown currents and voltages, 89

mesh-current analysis *(continued)*
  steps in, 86
  substituting device voltages into KVL
    equations, 88
  for three or more meshes, 92–94
  for two-mesh circuits, 90–92
  with voltage sources, 86–89
meshes, defined, 83
microelectromechanical systems
  (MEMS), 342
MOSFETs (metal-oxide semiconductor
  field-effect transistors), 142
Multisim software, 14

• N •

nanotechnology, 341–342
narrowband RLC series circuits, 329
National Instrument's Multisim
  software, 14
Ngspice software, 14
node voltage, 67–68, 76. *See also* node-
  voltage analysis (NVA)
nodes, 24, 30, 67
node-voltage analysis (NVA)
  of dependent sources, 137–138
  described, 11
  Kirchhoff's current law in, 69, 70, 75,
    78, 81
  linear algebra for, 73
  matrix form for equations, 72–73
  matrix software for, 73
  Ohm's law in, 70–72
  in phasor domain, 270–271
  in the *s*-domain, 303–304
  solving for unknown voltages with a
    current, 74–76
  solving for unknown voltages with a
    voltage source, 80–82
  steps in, 69
  for three or more node voltages, 76–79
noninverting op amps
  analysis of, 160–162
  in differential amplifier or subtractor,
    166–168
  voltage follower, 161, 162–163

Norton's theorem. *See also* Thévenin's
  theorem
  applying to circuit with multiple
    sources, 119–121
  delivering maximum power to a load
    circuit, 127–129
  described, 12
  finding Norton equivalent circuit from
    Thévenin equivalent, 122
  finding open-circuit voltage, 119–120
  finding short-circuit current, 120–121
  interface between source and load
    circuits, 114
  Norton current source, 116, 118,
    121, 126
  Norton equivalent in the phasor
    domain, 268
  Norton equivalent in the *s*-domain,
    309–311
  Norton resistance, 116, 122
  for RL parallel circuit with network of
    resistors, 225
  source circuit and load circuit in,
    113, 114
  source circuit equivalent using, 115
  source transformation with, 122
  superposition with, 124
NPN bipolar transistors, 146
NVA. *See* node-voltage analysis

• O •

octave log-frequency scales, 319
ohms, 17
Ohm's law
  applied to resistors, 16–17
  calculating device currents using,
    76, 77
  calculating power dissipated by
    resistors, 18
  for capacitance, 212
  defined, 16–17
  described, 11
  finding *i-v* characteristics using,
    43–44
  in mesh-current analysis, 87–88

in node-voltage analysis, 69, 70–72
phasor form for capacitors, 262
phasor form for inductors, 263
in terms of impedance, 260
open circuits
  combining with ideal switches,
    20–21
  defined, 20
  replacing current sources with, 99,
    102–103
  Thévenin equivalent using open-circuit
    loads, 116
  Thévenin voltage source, 116,
    117–118, 125
operational amplifiers (op amps)
  as active devices, 12
  capacitors in circuit with, 205–208
  circuit symbols, 156
  constraints in the s-domain, 299
  dependent source model, 158–159
  described, 12
  differential-equation-solving circuit,
    208–210
  differentiator circuit, 207–208
  digital-to-analog conversion with,
    172, 338
  equation solving using, 170–171
  equations for constraints, 159–160
  ideal, transfer characteristics of,
    157–158
  instrumentation amplifier, 168–169
  integrated circuits for, 155
  integrator as, 205–207, 210
  interface techniques using, 339
  inverting, 163–164
  inverting and noninverting input equal
    in, 161
  inverting and noninverting voltage
    equal in, 162
  linear region of, 157, 158
  negative saturated region of, 157, 158
  negative versus positive feedback
    for, 160
  noninverting, analysis, 160–162
  positive saturated region of, 157, 158

power of, 155
subtractor or differential amplifier,
  166–168
summer or summing amplifier,
  164–166, 172
systems built with, 171–172
terminals of, 156
uses for, 12
voltage division with, 161
voltage follower (noninverting), 161,
  162–163
voltage gain for, 157
voltage output of, 157

• P •

parallel circuits
  capacitance in, 199–200
  in Christmas lights, 39
  combining series and parallel
    resistors, 40
  conductance of, 37–38
  converting to and from series circuits,
    45–49
  current division in the s-domain,
    302–303
  described, 36
  equivalent resistor combinations,
    38–39
  finding currents with multiple voltage
    sources, 60–63
  finding the current through each device
    in, 59–60
  inductance in, 204–205
  RL, 212, 224–231
  RLC, 234, 246–251, 330–332
  starting analysis of, 36–37
  two resistors in parallel, 38
  using current division repeatedly for,
    63–64
  voltage division for multiple current
    sources and parallel resistors,
    52–55
  voltage uniform in, 31, 36

partial fraction expansion
  with complex poles, 278–280
  for damped sinusoids, 278
  finding transform pairs after, 279
  with multiple poles, 280–282
  steps in, 276–278
particular solution
  defined, 213
  for RC series circuit, 220–221
  for RL parallel circuit, 229
  for RLC parallel circuit, 251
  for RLC series circuit, 243–244
passive circuits, constant or gain for, 134
passive sign convention, 10
phase of frequencies, 314
phasor analysis
  differential equations not needed for,
    13, 255
  division techniques in phasor form,
    264–266
  mesh-current analysis in, 271–272
  node-voltage analysis in, 270–271
  Ohm's law in phasor form for
    capacitors, 262
  Ohm's law in phasor form for
    inductors, 263
  with sine wave input, 255
  steps in, 263–264
  superposition for phasor output,
    266–267
  Thévenin and Norton equivalents in,
    268–270
phasor diagrams, 261
phasors
  defined, 255
  forms of, 256–258
  properties of, 258–259
planar circuits, 83–84
PNP bipolar transistors, 146
polar form of phasors, 257
polarities, 9, 10
pole-zero diagrams, 282–283, 284
potentiometers, 337

power
  calculating dissipation by resistors, 18
  defined, 15
  delivering maximum to a load circuit,
    127–129
  dissipation by resistors, 18
  equation defining, 202
  finding for capacitors, 196–198
  positive versus negative, 15
  units of, 16
power supplies, schematic symbols
    for, 22
practical applications, 337–340
proportionality constant, 135, 260. *See
    also* gain
proportionality property of linearity,
    96–98. *See also* superposition
pulse
  impulse, 176–180
  ramp, 184
  rectangular, 176, 182, 184

• *Q* •

quality factor of band-pass filters, 329

• *R* •

ramp function, 182–184, 275, 284
RC series circuits
  band-pass filter using, 332–333
  band-reject filter using, 332–333
  capacitor output values for time
    constants, 224
  characteristic equation for, 218–219
  defined, 212
  element constraint for capacitor,
    216, 286
  finding the total response, 217, 222–224
  finding the voltage across the
    resistor, 215
  finding the zero-input response,
    217–219, 306–307

finding the zero-state response, 219–222, 306, 307–309
first-order differential equations for, 216, 217–218, 220, 221, 222
high-pass filter using, 325, 326
i-v characteristics for devices, 216, 286
Laplace transform for, 286–288
low-pass filter using, 325–326
Ohm's law for capacitor, 212
plotting capacitor voltage, 223
in the *s*-domain, 306–309
simple circuit, 215–217
for speaker systems, 327
state variables, 216
time constant for, 219
reactance (imaginary part of impedance), 260
rectangular form of phasors, 257
rectangular pulse, 176, 182, 184
reference point
  for circuits, 30
  for node voltages, 67
Remember icon, 4
resistance
  current-voltage characteristics with, 16
  defined, 17
  of ideal voltage source, 101–102
  infinite, 20
  input, for common collector circuit, 152–153
  Norton, 116, 122
  ohms as units of, 17
  in Ohm's law, 17
  in series circuits, 35–36
  of short circuits, 19
  of superconductors, 17
  Thévenin, 116, 118–119, 122, 126–127, 141–142
  total, conductance as, 37
  Wheatstone bridge circuit for measuring, 339, 340
  of wires connected at nodes, 24

resistors. *See also* current division; voltage division
  calculating power dissipated by, 18
  combining series and parallel, 40
  equivalent combinations in parallel circuits, 38–39
  in first-order circuits, 13, 211, 212
  impedance and admittance in the *s*-domain, 299
  impedance replacing, 264
  interface techniques using, 338–339
  i-v characteristics of, 16
  i-v relationships in the *s*-domain, 297–298
  linear, *i-v* characteristics of, 17
  Ohm's law applied to, 16–17
  in parallel, shortcut for, 38
  phasor diagram of, 261
  purposes of, 16
  in RC series circuits, 212
  in RL parallel circuits, 212
  in second-order circuits, 13
RL parallel circuits
  characteristic equation for, 227–228
  defined, 212
  element constraint for inductor, 225, 290
  finding the total response, 230–231
  finding the zero-input response, 226–228
  finding the zero-state response, 228–230
  first-order differential equations for, 225, 226, 227, 228–229
  i-v characteristics for devices, 225, 290
  i-v relationship for inductor, 212
  Laplace transform for, 290–292
  network of resistors with, 225
  resistor current in, 225
  simple circuit, 225–226
  time constant (*L/R*) for, 228, 231
RLC parallel circuits
  band-pass filter using, 330–332
  breaking into two circuits, 247
  capacitor current for, 265–266
  characteristic equation for, 249
  described, 246

RLC parallel circuits *(continued)*
  element constraint for inductor, 247
  example, 234
  finding the total response, 251
  finding the zero-input response, 249–250
  finding the zero-state response, 250–251
  homogeneous solution for, 251
  illustrating voltage division and parallel
    equivalence, 265–266
  particular solution for, 251
  second-order differential equations for,
    247, 248, 249–251
  setting up a typical circuit, 247–249
  using duality for, 246, 249, 250, 251
RLC series circuits
  band-pass filter using, 327–330
  band-reject filter using, 330
  breaking into two circuits, 237
  capacitor voltage for, 265
  characteristic equation for, 240
  described, 236–237
  element constraint for capacitor, 238,
    292–293
  element constraint for inductor, 238, 292
  example, 234
  finding the total response, 245–246
  finding the zero-input response,
    239–242
  finding the zero-state response,
    242–244
  homogeneous solution for, 243
  illustrating voltage division and series
    equivalence, 264–265
  Laplace transform for, 292–294
  narrowband versus wideband, 329
  particular solution for, 243–244
  plotting the zero-input response by
    resistance, 242
  second-order differential equations for,
    238, 239–240, 243–244
  simple circuit, 237–239
  step responses for values of
    resistances, 245–246
rotating vectors, phasors as, 256

• S •

s, frequency variable, 274
sawtooth function, 184
scalar multiplier. *See* gain
scale factor. *See* gain
schematics
  defined, 21
  gate symbols in, 21, 22
  loops in, 22–23
  nodes in, 24
  power supply symbols in, 22
  voltage rises and drops in, 27
  wire symbols in, 21, 22
s-domain. *See also* Laplace transform
  admittance in, 299
  connection constraints in, 296
  current division for parallel circuits in,
    302–303
  device constraints in, 297–299
  F(*s*) function in, 274
  impedance in, 299
  Laplace transforms as ratios of
    polynomials in, 282
  mesh-current analysis in, 304–305
  node-voltage analysis in, 303–304
  superposition in, 305–309
  Thévenin and Norton equivalents in,
    309–311
  voltage division with series circuits in,
    300–301
  zero-input transforms in, 306–307
  zero-state transforms in, 306, 307–309
second-order circuits. *See also specific
    types*
  described, 13, 233
  finding the total response, 234–235,
    245–246
  finding the zero-input response, 234,
    239–242
  finding the zero-state response, 234,
    242–244
  RLC parallel, 234, 246–251, 330–332

voltage sources, independent (continued)
relationship between dependent and
independent sources, 136
replacing with short circuits, 101–102
resistance of ideal source, 101–102
superposition with one voltage source
and one current source, 100–101,
107–108
superposition with three or more
sources, 108–111
superposition with two sources,
103–105
Thévenin equivalent for single
independent source, 117–119
Thévenin (open circuit), 116,
117–118, 120
turning on and off, 101–102, 136
voltage-controlled current source
(VCCS), 134–135, 297
voltage-controlled voltage source
(VCVS), 134–135, 297

• *W* •

Warning! icon, 4
watts, 16
weight factor. *See* gain
Wheatstone bridge circuit, 339, 340

wide bandgap semiconductors, 343–344
wideband RLC series circuits, 329
wires, 21, 22, 24
woofers, 327, 338

• *Z* •

zero-input response calculation
first-order circuits, 212
RC series circuits, 217–219
RL parallel circuits, 226–228
RLC parallel circuits, 249–250
RLC series circuits, 239–242
in the *s*-domain, 306–307
second-order circuits, 234
zero-state response calculation
defined, 219
homogeneous solution for, 213–214,
220, 228–229, 243, 251
particular solution for, 213, 220–221,
229, 243–244, 251
RC series circuits, 219–222
RL parallel circuits, 228–230
RLC parallel circuits, 250–251
RLC series circuits, 242–244
in the *s*-domain, 306, 307–309
second-order circuits, 234

RLC series, 234, 236–246, 292–294,
327–330
two storage elements in, 239
second-order differential equations
characteristic equation for, 236, 240
with constant coefficients, steps for
solving, 234
converting to algebraic equations, 236
exponential function with, 235
homogeneous solution for, 235, 243
particular solution for, 243–244
for RLC parallel circuits, 247, 248,
249–251
for RLC series circuits, 238, 239–240,
243–244
second-order circuits described
by, 233
semiconductors, wide bandgap,
343–344
series circuits
analyzing, 35–36
capacitance in, 200
combining series and parallel
resistors, 40
converting to and from parallel circuits,
45–49
current uniform in, 31, 34
defined, 34
inductance in, 203–204
RC, 212, 215–224
RLC, 234, 236–246, 292–294, 327–330
voltage division with, 49–52, 300–301
shifted cosine function, 257
short circuits
combining with ideal switches, 20–21
described, 19
i-v characteristics of, 18
Norton current source, 116, 118, 126
replacing voltage sources with, 98,
101–102
Thévenin equivalent using short-circuit
loads, 116
siemens, 17

signals
defined, 13
sinusoidal, 13, 255, 256
time-varying, 13
Simulation Program for Integrated Circuit
Emphasis (SPICE) software, 14
sine function, 275, 284, 285
sinusoidal functions
advanced, 187–188
amplitude of, 186
connecting to exponentials via Euler's
formula, 190–191
delayed, 187–188
expanding, 189–190
Fourier coefficients of, 190
frequency of, 186
period of, 186
phase shift of, 187–189
time-shifted, 186–187
sinusoidal signals, 13, 255, 256
smartphone touchscreens, 341
software, 14, 73
source circuit. *See also* Norton's
theorem; Thévenin's theorem
defined, 113, 114
Norton equivalent for, 115
Thévenin equivalent for, 114–115, 116
source lead of JFET transistor, 143
source transformation
constraining equation for, 45
converting current source to voltage
source, 47–49
converting voltage source to current
source, 45–47
with dependent sources, 138–140
finding equivalent circuits, 42–44
Norton's theorem with, 122
Thévenin's theorem with, 122–124
speaker systems, 327, 338
SPICE (Simulation Program for Integrated
Circuit Emphasis) software, 14
state variables, 216, 239
step function, 177, 180–184, 275

stud finders, electronic, 340
subtractor or differential amplifier, 166–168
summer or summing amplifier, 164–166, 172, 210
supercapacitors, 343
superconductors, 17, 343
superposition
    described, 12, 95
    diagram of, 99
    Norton's theorem with, 124
    with one voltage source and one current source, 100–101, 107–108
    for phasor output, 266–267
    proportionality concept for, 96–98
    removing a current source, 102–103
    removing a voltage source, 101–102
    in the s-domain, 305–309
    steps for applying, 98–99
    Thévenin's theorem with, 124–127
    with three or more sources, 108–111
    with two current sources, 105–106
    with two voltage sources, 103–105

● T ●

Technical Stuff icon, 2, 4
technologies affecting circuits, 341–344
terminals
    of bipolar transistors, 145
    of op amps, 156
Thévenin's theorem. See also Norton's theorem
    applying to circuit with multiple sources, 122–124
    applying to circuit with one voltage source, 117–119
    delivering maximum power to a load circuit, 127–129
    described, 12
    equivalent using open-circuit loads, 116
    equivalent using short-circuit loads, 116
    finding Norton equivalent circuit from Thévenin equivalent, 122

interface between source and load circuits, 114
source circuit and load circuit in, 113, 114
source circuit equivalent using, 114–115
source transformation with, 122–124
superposition with, 124–127
Thévenin equivalent in phasor domain, 268–270
Thévenin equivalent in the s-domain, 309–311
Thévenin resistance, 116, 118–119, 122, 126–127, 141–142
Thévenin voltage source, 116, 117–118, 120, 125, 141, 142
time
    electrical power as function of, 15
    variable for, 8
time constant, 219, 228, 231
time derivative of phasors, 259
timer circuits, 340
time-shifted exponential function, 185
time-shifted sinusoidal functions, 186–187
time-shifted step function, 181, 182
time-varying signals, 13, 175. See also capacitors; first-order circuits; functions; inductors; second-order circuits
timing circuits, 16
Tip icon, 4
total response calculation
    first-order circuits, 213
    RC series circuits, 217, 222–224
    RL parallel circuits, 230–231
    RLC parallel circuits, 251
    RLC series circuits, 245–246
    second-order circuits, 234–235
touchscreens, smartphone, 341
transconductance of voltage-controlled current source, 135
transducer, 339

transfer function
    Bode plots related to characteristics of, 320–324
    as complex number, 314
    defined, 314
    gain and phase relationships, 314
    for RC high-pass filter, 326
    for RC low-pass filter, 325, 326
    for RLC parallel band-pass filter, 330–332
    for RLC series band-pass filter, 328
    for RLC series band-reject filter, 330
transformation techniques. See Laplace transform; source transformation
transistors
    bipolar, 145–153
    described, 12, 142
    field-effect (FET), 142
    origin of name, 133
transresistance of CCVS, 135
triangle function, 183–184
turning on and off
    current sources, 102–103, 136
    voltage sources, 101–102, 136
tweeters, 327, 338

● V ●

variables, 1, 9
VCCS (voltage-controlled current source), 134–135, 297
VCVS (voltage-controlled voltage source), 134–135, 297
voltage. See also Kirchhoff's voltage law (KVL)
    as an across variable, 9
    birds on high-voltage lines, 19
    calculating for devices, 68
    of capacitors, relationship of current to, 195–196
    connection equations for, 32
    device equations for, 31, 33–34
    device, in mesh-current analysis, 88

drops, 27
in electrical power function, 15
finding with Kirchhoff's laws, 32
ground as zero voltage point, 9
joules per coulomb as units of, 16
in Ohm's law, 17
in parallel circuits, 31
positive versus negative, 10
reduced by resistors, 16
rises, 26–27
in series circuits, 35
two points needed to measure, 9
voltage division
    maximum power solution using, 128
    for multiple current sources and parallel resistors, 52–55
    with op amps, 161
    in phasor form, 264–266
    as proportionality concept, 97
    for RC high-pass filter, 326
    for RC low-pass filter, 325
    for RLC series band-pass filter, 327
    for RLC series band-reject filter, 330
    for series circuit in the s-domain, 300–301
    for series circuit with multiple resistors, 51–52
    for series circuit with one resistor, 49–51
    using repeatedly, 55–56
voltage follower op amp, 161, 162–163
voltage rises and drops. See Kirchhoff's voltage law (KVL)
voltage sources, independent. See dependent sources
    batteries as, 19
    converting to and from current sources, 45–49
    finding the current through each in parallel circuits, 59–60
    mesh-current analysis with, 86–87
    multiple, current division for, 60
    node-voltage analysis for, 80–82

Apple & Mac

iPad For Dummies,
5th Edition
978-1-118-49823-1

iPhone 5 For Dummies,
6th Edition
978-1-118-35201-4

MacBook For Dummies,
4th Edition
978-1-118-20920-2

OS X Mountain Lion
For Dummies
978-1-118-39418-2

Blogging & Social Media

Facebook For Dummies,
4th Edition
978-1-118-09562-1

Mom Blogging
For Dummies
978-1-118-03843-7

Pinterest For Dummies
978-1-118-32800-2

WordPress For Dummies,
5th Edition
978-1-118-38318-6

Business

Commodities For Dummies,
2nd Edition
978-1-118-01687-9

Investing For Dummies,
6th Edition
978-0-470-90545-6

Personal Finance
For Dummies,
7th Edition
978-1-118-11785-9

QuickBooks 2013
For Dummies
978-1-118-35641-8

Small Business Marketing Kit
For Dummies,
3rd Edition
978-1-118-31183-7

Careers

Job Interviews
For Dummies,
4th Edition
978-1-118-11290-8

Job Searching with
Social Media
For Dummies
978-0-470-93072-4

Personal Branding
For Dummies
978-1-118-11792-7

Resumes For Dummies,
6th Edition
978-0-470-87361-8

Success as a Mediator
For Dummies
978-1-118-07862-4

Diet & Nutrition

Belly Fat Diet For Dummies
978-1-118-34585-6

Eating Clean For Dummies
978-1-118-00013-7

Nutrition For Dummies,
5th Edition
978-0-470-93231-5

Digital Photography

Digital Photography
For Dummies,
7th Edition
978-1-118-09203-3

Digital SLR Cameras &
Photography For Dummies,
4th Edition
978-1-118-14489-3

Photoshop Elements 11
For Dummies
978-1-118-40821-6

Gardening

Herb Gardening
For Dummies,
2nd Edition
978-0-470-61778-6

Vegetable Gardening
For Dummies,
2nd Edition
978-0-470-49870-5

Health

Anti-Inflammation Diet
For Dummies
978-1-118-02381-5

Diabetes For Dummies,
3rd Edition
978-0-470-27086-8

Living Paleo For Dummies
978-1-118-29405-5

Hobbies

Beekeeping
For Dummies
978-0-470-43065-1

eBay For Dummies,
7th Edition
978-1-118-09806-6

Raising Chickens
For Dummies
978-0-470-46544-8

Wine For Dummies,
5th Edition
978-1-118-28872-6

Writing Young Adult Fiction
For Dummies
978-0-470-94954-2

Language &
Foreign Language

500 Spanish Verbs
For Dummies
978-1-118-02382-2

English Grammar
For Dummies,
2nd Edition
978-0-470-54664-2

French All-in One
For Dummies
978-1-118-22815-9

German Essentials
For Dummies
978-1-118-18422-6

Italian For Dummies
2nd Edition
978-1-118-00465-4

 Available in print and e-book formats.

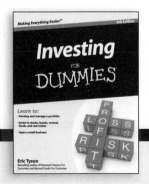
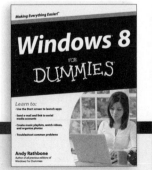

Math & Science

Algebra I For Dummies,
2nd Edition
978-0-470-55964-2

Anatomy and Physiology
For Dummies,
2nd Edition
978-0-470-92326-9

Astronomy For Dummies,
3rd Edition
978-1-118-37697-3

Biology For Dummies,
2nd Edition
978-0-470-59875-7

Chemistry For Dummies,
2nd Edition
978-1-1180-0730-3

Pre-Algebra Essentials
For Dummies
978-0-470-61838-7

Microsoft Office

Excel 2013 For Dummies
978-1-118-51012-4

Office 2013 All-in-One
For Dummies
978-1-118-51636-2

PowerPoint 2013
For Dummies
978-1-118-50253-2

Word 2013 For Dummies
978-1-118-49123-2

Music

Blues Harmonica
For Dummies
978-1-118-25269-7

Guitar For Dummies,
3rd Edition
978-1-118-11554-1

iPod & iTunes
For Dummies,
10th Edition
978-1-118-50864-0

Programming

Android Application
Development For
Dummies, 2nd Edition
978-1-118-38710-8

iOS 6 Application
Development For Dummies
978-1-118-50880-0

Java For Dummies,
5th Edition
978-0-470-37173-2

Religion & Inspiration

The Bible For Dummies
978-0-7645-5296-0

Buddhism For Dummies,
2nd Edition
978-1-118-02379-2

Catholicism For Dummies,
2nd Edition
978-1-118-07778-8

Self-Help & Relationships

Bipolar Disorder
For Dummies,
2nd Edition
978-1-118-33882-7

Meditation For Dummies,
3rd Edition
978-1-118-29144-3

Seniors

Computers For Seniors
For Dummies,
3rd Edition
978-1-118-11553-4

iPad For Seniors
For Dummies,
5th Edition
978-1-118-49708-1

Social Security
For Dummies
978-1-118-20573-0

Smartphones & Tablets

Android Phones
For Dummies
978-1-118-16952-0

Kindle Fire HD
For Dummies
978-1-118-42223-6

NOOK HD For Dummies,
Portable Edition
978-1-118-39498-4

Surface For Dummies
978-1-118-49634-3

Test Prep

ACT For Dummies,
5th Edition
978-1-118-01259-8

ASVAB For Dummies,
3rd Edition
978-0-470-63760-9

GRE For Dummies,
7th Edition
978-0-470-88921-3

Officer Candidate Tests,
For Dummies
978-0-470-59876-4

Physician's Assistant Exam
For Dummies
978-1-118-11556-5

Series 7 Exam
For Dummies
978-0-470-09932-2

Windows 8

Windows 8 For Dummies
978-1-118-13461-0

Windows 8 For Dummies,
Book + DVD Bundle
978-1-118-27167-4

Windows 8 All-in-One
For Dummies
978-1-118-11920-4

Available in print and e-book formats.

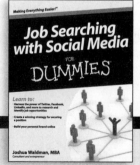

Take Dummies with you everywhere you go!

Whether you're excited about e-books, want more from the web, must have your mobile apps, or swept up in social media, Dummies makes everything easier .